BILINGUAL LECTURES OF DESIGN AND CULTURE

设计文化
双语教程

应宜文 编著

解析国内外设计文化理论与设计发展史，通过前沿经典的设计范例及其设计思想、方法和流程以启迪设计思维，培养读者对设计文献的双语理解能力与提高设计自主创新能力，融理论性、实用性和前瞻性于一体

中国建筑工业出版社

图书在版编目（CIP）数据

设计文化双语教程/应宜文编著.—北京：中国建筑工业出版社，2017.12

ISBN 978-7-112-21444-0

Ⅰ.①设… Ⅱ.①应… Ⅲ.①建筑设计-双语教学-高等学校-教材 Ⅳ.①TU2

中国版本图书馆CIP数据核字（2017）第265789号

本教程旨在为高校艺术与设计学专业本科及研究生提供一本解析国内外设计文化理论与设计发展史，通过前沿经典的设计范例及其设计思想、方法和流程以启迪设计思维，培养读者对设计文献的双语理解能力与提高设计自主创新能力，融理论性、实用性和前瞻性于一体的教学用书。本教程理论与实例相辅相成，第1至5章阐释设计理论，第6~10章解读设计范例，"附录一：中国古典设计文献分类辑录"围绕"设计文化"要义对现存古籍文献进行详细梳理与分类，从我国优秀的传统文化中汲取设计灵感；"附录二：西方艺术设计文献分类辑录"精选具有代表意义的西方艺术设计典籍并分类。

本教程可作为高等院校艺术设计学科中环境设计、景观设计、视觉传达设计、工业设计、新媒体设计等专业的双语教学用书，也可供其他艺术设计相关专业参考使用，亦可作为读者今后深化研究、开拓设计构思及创作的参考书。

责任编辑：吴　绫　李东禧
责任校对：王　瑞　焦　乐

设计文化双语教程
应宜文　编著
*
中国建筑工业出版社出版、发行（北京海淀三里河路9号）
各地新华书店、建筑书店经销
北京嘉泰利德公司制版
大厂回族自治县正兴印务有限公司印刷
*
开本：787×1092毫米　1/16　印张：$12\frac{3}{4}$　字数：203千字
2017年12月第一版　2017年12月第一次印刷
定价：48.00元（赠课件）
ISBN 978-7-112-21444-0
　　　（31127）

版权所有　翻印必究
如有印装质量问题，可寄本社退换
（邮政编码 100037）

本书为 2016 年度浙江工业大学重点建设教材项目教研成果

（项目批准号 JC1617）

前 言
Preface

　　从丰富多彩的当代设计中，我们发现追求新奇新异效果的设计偏多而体现中国文化的设计甚少，如何来改善这种现象？我们需要弥补设计作品缺少文化内涵的不足，以中国传统文化作为基石，积极探索与开创具有高层次文化品位的设计作品。未来中国设计的趋向不是模仿西方设计，而是自主创新研发蕴含中国文化特色的原创设计。21世纪优秀的设计人才具有传承我国历史文脉与弘扬民族文化特色的使命，必定是具备"跨学科、重文化、多领域"的综合型创新人才，顺应设计国际化的大趋势，拓展原创中国设计屹立于世界设计之林。

　　本教程主要是为高校艺术与设计学专业的本科生及研究生提供一本解读国内外设计文化理论与设计发展史，通过新颖前沿的高水平设计范例以启迪设计思维，培养读者对设计文献、设计作品的双语理解能力并提高设计自主创新能力，融理论性、实用性和前瞻性于一体的教学用书。本教程的"附录一：中国古典设计文献分类辑录"，围绕"设计文化"的要义对现存古籍文献进行较为详细的梳理，并分为中国古代设计文化思想文献、中国古代设计文明成就文献、中国古代设计营造方法文献三类，有助于读者深入研究中国古代造物思想，从中国古典设计文化中孕育设计创造灵感；"附录二：西方艺术设计文献分类辑录"，精选具有代表意义的西方艺术设计书籍，并分为西方设计教育文献、西方设计文化与设计理论文献、西方设计作品专辑三类，可作为读者今后持续深化研究、开拓设计构思以及设计创作的参考资料。使用本教材，不仅可以提高读者理解专业文献、解读国内外设计的综合能力，还能够使读者从我国传统文化的视角开阔设计视野及创新思路，为其日后从事相关的设计创意工作和理论研究打下坚实的基础。

　　本教程主题明确、理论与实例兼顾、知识覆盖面广、语言规范、难度适中，有益于启发创造性思维，有益于提高学生理解设计文化双语文献和解读设计作品的能力。本教材适合于艺术与设计学二年级以上学生使用，建议采用32课时或48课时教学。本教材的特色在于弘扬我国古代"造物"思想与唯物求新观念，"造物"思想正是当代"设计"的原点，注重富有中国文化特色的设计内涵以提高综合创新能力等；精选任天、朱昱宁、郑昱、俞雪莱等设计师和吕勤智教授的经典创意设计实例，书中既有真实而富有艺术感染力的设计范例，又有生动、准确的中英双语解析与导读，加强艺术设计专业用语理解与运用，集思广益、图文并茂、内容新颖，能够使读者对设计学理论有一个系统全面的认识，并能够使读者获取最新设计文化知识与信息。

本教程在撰写中，得到了中国艺术研究院任平教授、GEORGIAN COURT UNIVERSITY 美国乔治亚大学 Lili Bruess 教授、设计研究所的著名设计师们的支持与帮助。诚挚感谢毕业于美国哈佛大学，现为中国美术学院教师任天博士在教程撰写中提出了很多宝贵的建议与指导。中国建筑工业出版社的吴绫编辑在全书出版过程中给予了笔者中肯的建议。在此，笔者向他们表示由衷的感谢！

由于笔者水平所限，教材中若有误漏欠妥之处，恳请读者指正。

本教材为 2016 年度浙江工业大学重点教材建设项目（JC1617）的教研成果。

目 录

第1章 导论 //// 1
1.1 文化与文明不同语境的释词考辨 //// 2
1.2 "设计"释词与古今考辨 //// 6

第2章 中国古典设计文献选读 //// 9
2.1 宋·郭熙《林泉高致》(节选) //// 10
2.2 元·《大元毡罽工物记·御用》(节选) //// 13
2.3 明·宋应星《天工开物》(节选) //// 15
2.4 明·计成《园冶·装折》(节选) //// 18
2.5 清·李斗《工段营造录》(节选) //// 21
2.6 清·寂园叟《匋雅》(节选) //// 24
2.7 清·李渔《闲情偶寄·器玩部》(节选) //// 26
2.8 清·钱泳《履园丛话·造园》(节选) //// 29

第3章 中国近代设计文化概论 //// 33
3.1 先导:中国设计启蒙的教育家 //// 34
3.2 先锋:中西合璧的设计教材 //// 51
3.3 引鉴:西方文化影响下的中国设计 //// 64

第4章 西方古典时代至18世纪设计文化选读 //// 67
4.1 西方古典时代的设计文化 //// 68
4.2 西方中世纪的设计文化 //// 71

Contents

Chapter One : Introduction ／／／ 1

    1.1 Textual Research on Culture and Civilization ／／／ 2

    1.2 Ancient and Modern Textual Research on Design ／／／ 6

Chapter Two : Selected Readings in Classical Chinese Design Text ／／／ 9

    2.1 *Lin Quan Gao Zhi*. Author Guo Xi. Song Dynasty (Selected Sections) ／／／ 10

    2.2 *Da Yuan Zhan Ji Gong Wu Ji · Yu Yong*. Yuan Dynasty (Selected Sections) ／／／ 13

    2.3 *Tian Gong Kai Wu*. Author Song Yingxing. Ming Dynasty (Selected Sections) ／／／ 15

    2.4 *Yuan Ye·Zhuang Zhe*. Author Ji Cheng. Ming Dynasty (Selected Sections) ／／／ 18

    2.5 *Gong Duan Ying Zao Lu*. Author Li Dou. Qing Dynasty (Selected Sections) ／／／ 21

    2.6 *Tao Ya*. Author Ji Yuansou. Qing Dynasty (Selected Sections) ／／／ 24

    2.7 *Xian Qing Ou Ji*. Author Li Yu. Qing Dynasty (Selected Sections) ／／／ 26

    2.8 *Lv Yuan Cong Hua*. Author Qian Yong. Qing Dynasty (Selected Sections) ／／／ 29

Chapter Three : Topic on Chinese Design Culture in Modern Times ／／／ 33

    3.1 Forerunner: Chinese Enlightened Design Educator ／／／ 34

    3.2 Pioneer: Design Textbooks Combining of Chinese and Western Styles ／／／ 51

    3.3 Import: Chinese Design under the Influence of Western Culture ／／／ 64

Chapter Four : Selected Readings of Western Design Culture from Classical Era to 18th Century ／／／ 67

    4.1 Design Culture of Western Classical Era ／／／ 68

    4.2 Design Culture of Western Middle Ages ／／／ 71

4.3 西方文艺复兴的设计文化 /// 73

4.4 巴洛克时期、洛可可时期的设计文化 /// 76

第 5 章　西方 19 世纪至 20 世纪设计文化选读 /// 79

5.1 工业社会早期的设计文化 /// 80

5.2 19 世纪末至 20 世纪初期西方设计思潮：工艺美术运动 /// 83

5.3 20 世纪初期西方设计思潮：新艺术运动 /// 90

5.4 20 世纪西方建筑文化思潮：芝加哥学派 /// 98

第 6 章　新农村改造设计 /// 105

6.1 任天设计师的心愿：用设计播下一颗种子 /// 106

6.2 任天设计师作品实例：上横街村 /// 110

6.3 传统文化与设计创新 /// 113

6.4 设计理念与设计方法 /// 116

第 7 章　建筑装置设计 /// 123

7.1 任天设计师作品实例：住棚设计 /// 124

7.2 创意构思与设计方法：住棚设计（图 7-4~ 图 7-15）/// 126

7.3 任天设计师作品实例：记忆山丘 /// 130

7.4 设计理念与设计方法：记忆山丘 /// 132

4.3　Design Culture of Western Renaissance ／／／ 73

4.4　Design Culture of Baroque and Rococo Period ／／／ 76

Chapter Five：Selected Readings of Western Design Culture from 19th to 20th Century ／／／ 79

5.1　Design Culture of Early Industrial Society ／／／ 80

5.2　Trend of Western Design Thoughts from the Late 19th Century to the Early 20th Century：Arts & Crafts Movement ／／／ 83

5.3　Trend of Western Design Thoughts from the Early 20th Century: Arts Nouveau ／／／ 90

5.4　Trend of Western Architecture Design Thoughts from the 20th Century: Chicago School ／／／ 98

Chapter Six：Design of New Rural Reconstruction ／／／ 105

6.1　Cherished Desire of Designer Ren Tian: Sow the Seed of Hope with Design ／／／ 106

6.2　Work of Ren Tian Designer: Shang Heng Jie Village ／／／ 110

6.3　Traditional Culture and Design Innovation ／／／ 113

6.4　Design Ideas and Design Methods ／／／ 116

Chapter Seven：Building Device Design ／／／ 123

7.1　Work of Ren Tian Designer: SUKKAN Design ／／／ 124

7.2　Creative Ideas and Design Methods: SUKKAN Design (Illustration 7-4 to Illustration 7-15) ／／／ 126

7.3　Work of Ren Tian Designer: Memory Scape ／／／ 130

7.4　Design Concepts and Design Methods: Memory Scape ／／／ 132

第8章 工业产品设计 /// 135

    8.1 朱昱宁设计师的感言：设计传承文化 /// 136

    8.2 朱昱宁设计师作品实例：八宝吉祥扇 /// 137

    8.3 郑昱设计师的心得：设计源于生活 /// 140

    8.4 郑昱设计师作品实例：延绵·书桌设计 /// 141

    8.5 郑昱设计师作品实例：蒸汽加热餐具组合 /// 146

第9章 视觉文化设计 /// 151

    9.1 俞雪莱设计师的心得：设计需要良好的沟通 /// 152

    9.2 俞雪莱设计师作品实例：甲骨文的设计元素 /// 154

    9.3 俞雪莱设计师作品实例：中国结的设计元素 /// 156

    9.4 俞雪莱设计师作品实例：中国泥塑的设计元素 /// 157

第10章 历史建筑保护与再利用 /// 159

    10.1 吕勤智设计师的体会：设计引领生活 /// 160

    10.2 吕勤智设计师的作品实例：老土木楼建筑保护与再利用 /// 162

    10.3 传承文化再创辉煌：获奖设计作品赏析 /// 164

附录一：中国古典设计文献分类辑录 /// 167

附录二：西方艺术设计文献分类辑录 /// 172

征引及参考文献 /// 175

后　记 /// 179

目录
Contents

Chapter Eight : Industry Production Design /// 135

8.1 Understanding of Designer Zhu Yuning: Design Inherits Culture /// 136

8.2 Work of Zhu Yuning Designer: Eight Auspicious Symbols Fan /// 137

8.3 Feeling of Designer Zheng Yu: Design Comes from Life /// 140

8.4 Work of Zheng Yu Designer: Stretch Long & Continue Forever Desk Design /// 141

8.5 Work of Zheng Yu Designer: Half & Half /// 146

Chapter Nine : Visual Culture Design /// 151

9.1 Feeling of Shirley Yu Designer: Design Requires Good Communication /// 152

9.2 Work of Shirley Yu Designer: Design Element of Inscription on Bones or Tortoise Shells of the Shang Dynasty /// 154

9.3 Work of Shirley Yu Designer: Design Element of Chinese Knot /// 156

9.4 Work of Shirley Yu Designer: Design Element of Chinese Clay Sculpture /// 157

Chapter Ten : Historical Architecture Protection and Design /// 159

10.1 Learning from Lv Qinzhi Designer: Design Eagerly Looks forward to Life /// 160

10.2 Work of Lv Qinzhi Designer: Old Civil Construction Protection and Reuse Design /// 162

10.3 Inheriting Culture and Splendid Civilization: Appreciation of Prize-winning Design Work /// 164

The First Appendix: The Compilation of Chinese Classical Design Texts /// 167

The Second Appendix: The Compilation of Western Art Design Texts /// 172

References /// 175

Postscript /// 179

图版目录
List of Figures

第 1 章图：

图 1–1　汉·刘向《说苑》封面（见：四部丛刊景明钞本。或见：清光绪元年（1875 年）湖北崇文书局刻本）

图 1–2　汉·刘向《说苑》正文页（见：四部丛刊景明钞本。或见：清光绪元年（1875 年）湖北崇文书局刻本）

图 1–3　近代赵尔巽撰《清史稿》封面，第一百九十二卷
　　　　（见：1928 年清史馆本。或见：1942 年联合书店刻本）

图 1–4　近代赵尔巽撰《清史稿》第一百九十二卷
　　　　（见：1928 年清史馆本。或见：1942 年联合书店刻本）

图 1–5　周朝卜商撰《子夏易传》卷一封面（见：清通志堂经解本）

图 1–6　周朝卜商撰《子夏易传》卷一（见：清通志堂经解本）

图 1–7　宋代蔡沈撰《书经集传》封面（见：清文渊阁四库全书本）

图 1–8　宋代蔡沈撰《书经集传》（见：清文渊阁四库全书本）

第 2 章图：

图 2–1　宋·郭熙《林泉高致》，明代读书坊刻本，封面

图 2–2　宋·郭熙《林泉高致》，明代读书坊刻本，第 1–2 页

图 2–3　元·佚名《大元毡罽工物记·御用》，1916 年，上海仓圣明智大学铅印本，封面

图 2–4　元·佚名《大元毡罽工物记·御用》，1916 年，上海仓圣明智大学铅印本，第 3–4 页

图 2–5　明·宋应星《天工开物》，1930 年，上海华通书局影印本，封面

图 2–6　明·宋应星《天工开物》，1930 年，上海华通书局影印本，目录第 1 页

图 2-7　明·宋应星《天工开物》，1930 年，上海华通书局影印本，目录第 2 页

图 2-8　明·宋应星《天工开物》，1930 年，上海华通书局影印本，自序第 1 页

图 2-9　明·宋应星《天工开物》，1930 年，上海华通书局影印本，自序第 2 页

图 2-10　中国园林建筑的牛腿（手绘创作者：陈炜副教授，浙江工业大学设计艺术学院）

图 2-11　中国园林建筑的牛腿（手绘创作者：陈炜副教授，浙江工业大学设计艺术学院）

图 2-12　明·计成《园冶》三卷，1932 年，中国营造学社铅印本，封面

图 2-13　古建筑的斗栱（手绘作者：陈炜副教授，浙江工业大学设计艺术学院）

图 2-14　古建筑的牛腿（与斗栱相连部分）（手绘作者：陈炜副教授，浙江工业大学设计艺术学院）

图 2-15　清·李斗《工段营造录》，1931 年，中国营造学社铅印本，封面

图 2-16　清·李斗《工段营造录》，1931 年，中国营造学社铅印本，正文页

图 2-17　清·李斗《工段营造录》，1931 年，中国营造学社铅印本，斗栱，正文页

图 2-18　清·李斗《工段营造录》，1931 年，中国营造学社铅印本，封底版权页

图 2-19　清·寂园叟《匋雅》，1918 年，静园刻本，自序页

图 2-20　清·寂园叟《匋雅》，1918 年，静园刻本，第 1 页

图 2-21　清·李渔《闲情偶寄·器玩部》，清康熙十年（1671）刻本，封面

图 2-22　清·李渔《闲情偶寄·器玩部》，清康熙十年（1671）刻本，目录第 1 页

图 2-23　清·李渔《闲情偶寄·器玩部》，清康熙十年（1671）刻本，目录第 2 页

图 2-24　清·李渔《闲情偶寄·器玩部》，清康熙十年（1671）刻本，目录第 3 页

图 2-25　清·李渔《闲情偶寄·器玩部》，清康熙十年（1671）刻本，目录第 4 页

图 2-26　清·李渔《闲情偶寄·器玩部》，清康熙十年（1671）刻本，正文页

图 2-27　清·钱泳《履园丛话·造园》，清道光五年（1825 年）述德堂刻本，封面

图 2-28　清·钱泳《履园丛话·造园》，清道光五年（1825 年）述德堂刻本，目录第 1 页

图 2-29　清·钱泳《履园丛话·造园》，清道光五年（1825 年）述德堂刻本，目录第 2 页

图 2-30　清·钱泳《履园丛话·造园》，清道光五年（1825 年）述德堂刻本，正文页

第 3 章图：

图 3-1　陈之佛设计的《东方杂志》封面，第二十四卷，第十一号，1927 年 6 月 10 日出刊

图 3-2　陈之佛设计的《东方杂志》封面，第二十四卷，第十二号，1927 年 6 月 25 日出刊

图 3-3　1917 年鲁迅设计的北京大学校徽

图 3-4　陶元庆书籍封面设计《幻象的残象》之一

图 3-5　陶元庆书籍封面设计《幻象的残象》之二

图 3-6　陶元庆书籍封面设计《工人绥惠略夫》

图 3-7　陶元庆书籍封面设计《回家》

图 3-8　《钱君匋印存》封面

图 3-9　《钱君匋印存》书籍第 20 页篆刻作品

图 3-10　《钱君匋印存》书籍第 21 页篆刻作品

图 3-11　《钱君匋印存》书籍第 32 页篆刻作品

图 3-12　钱君匋编著《图案字文集》中"进行曲选"字体设计

图 3-13　钱君匋编著《图案字文集》中"高尚趣味"字体设计

图 3-14　钱君匋著《西洋古代美术史》，永祥印书馆，民国 35 年（1946 年）六月初版

图 3-15　钱君匋著《西洋近代美术史》，永祥印书馆，民国 35 年（1946 年）九月初版

图 3-16　陈之佛编著《图案教材》，上海：天马书店，1935 年，封面

图 3-17　陈之佛编著《图案教材》，上海：天马书店，1935 年，版权页

图 3-18　雷圭元著《工艺美术技法讲话》，南京：正中书局，1936 年

图 3-19　俞剑华编《最新立体图案法》，上海：商务印书馆，1929 年

图 3-20　郑棣、方炳潮撰编《图案教材》，浙江正楷印书局，1936 年初版，目录

图 3-21　郑棣、方炳潮撰《图案教材》，浙江正楷印书局，1936 年初版，例言

图 3-22　傅抱石编译《基本图案学》（职业学校教科书），民国 25 年（1936 年）2 月初版

图 3-23　傅抱石编译《基本工艺图案法》，商务印书馆，1939 年 3 月初版，封面

图 3-24　傅抱石编译《基本工艺图案法》，商务印书馆，1939 年 3 月初版，凡例

图 3-25　傅抱石编译《基本工艺图案法》，商务印书馆，1939 年 3 月初版，封底

图 3-26　茅剑青编《普通平面图案画法》，上海：北新书局，1936 年，封面

图 3-27　茅剑青编《普通平面图案画法》，上海：北新书局，1936 年，版权页

图 3-28　陆旋编《实用图案画法》，上海：新民图书馆，1921 年，封面

图 3-29　陆旋编《实用图案画法》，上海：新民图书馆，1921 年，版权页

图 3-30　朱西一主编《图案画法》，上海：中华书局，1947 年，封面

图 3-31　朱西一主编《图案画法》，上海：中华书局，1947 年，版权页

图 3-32　楼子尘原著《现代工艺图案构成法》，上海形象艺术社，1933 年，封面

图 3-33　楼子尘原著《现代工艺图案构成法》，上海形象艺术社，1933 年，自序

图版目录
List of Figures

图 3-34　楼子尘原著《现代工艺图案构成法》,上海形象艺术社,1933 年,版权页

图 3-35　上海发行的商业海报《沪景开彩图》,1896 年,中国最早的广告画

第 4 章图：

图 4-1　陶立克柱式(Doric order)(手绘表达：应宜文)

图 4-2　爱奥尼柱式(Ionic order)(手绘表达：应宜文)

图 4-3　古罗马万神庙建筑正门(手绘写生：应宜文)

图 4-4　古罗马万神庙建筑室内圆顶

图 4-5　意大利阿雷佐(Arezzo)生产的阿雷陶瓷(Arretine Ware)

图 4-6　波特兰花瓶(Portland Vase)(手绘表达：应宜文)

图 4-7　克里斯莫斯椅子(Klismos)(手绘表达：应宜文)

图 4-8　哥特式建筑顶部尖拱结构(ogival arch)(手绘表达：应宜文)

图 4-9　欧洲哥特式建筑写生(手绘表达：应宜文)

图 4-10　受哥特式建筑文化影响下"尖锐修长"风格的时尚服装设计(手绘表达：应宜文)

图 4-11　文艺复兴时期的欧洲圆顶建筑写生之一(手绘表达：应宜文)

图 4-12　文艺复兴时期的欧洲圆顶建筑写生之二(手绘表达：应宜文)

图 4-13　佛罗伦萨洗礼堂的青铜门(实地拍摄：王颖)

图 4-14　佛罗伦萨洗礼堂的青铜门装饰细节图(实地拍摄：王颖)

图 4-15　纪念意大利著名诗人但丁而设计的"但丁椅",这款椅子造型优雅均衡在当时很受欢迎

图 4-16　法国凡尔赛宫入口处写生(手绘表达：应宜文)

第 5 章图：

图 5-1　韦奇伍德研制的皇后陶瓷(摄影：王颖)

图 5-2　韦奇伍德研制生产的仿制古典风格的波特兰花瓶(邵宏主编,颜勇、黄虹等编著《西方设计：一部为生活制作艺术的历史》,长沙：湖南科学技术出版社,第 154 页)

图 5-3　英国皇家园艺总监帕克斯顿(Joseph Paxton)设计的博览会"水晶宫"(Crystal Palace)建筑(鲁石著《你应该读懂的 100 处世界建筑》,西安：陕西师范大学出

版社，第 266 页）

图 5-4　查尔斯·巴里（Charles Barry）与普金（Augustus Welby Northmore Pugin）两位建筑师合作设计的英国议会大厦（鲁石著《你应该读懂的 100 处世界建筑》，西安：陕西师范大学出版社，第 253 页）

图 5-5　1859 年威廉·莫里斯在乌普顿的著名建筑"红屋"（Red House），由菲利普·韦伯（Philips Webb）担任建筑平面设计（鲁石著《你应该读懂的 100 处世界建筑》，西安：陕西师范大学出版社，第 286 页）

图 5-6　由威廉·莫里斯设计的可以调节椅背的"莫里斯椅"（邵宏主编，颜勇、黄虹等编著《西方设计：一部为生活制作艺术的历史》，长沙：湖南科学技术出版社，第 199 页）

图 5-7　由理查德·诺曼·肖（Richard Norman Shaw）设计的伦敦老天鹅住宅建筑外立面（邵宏主编，颜勇、黄虹等编著《西方设计：一部为生活制作艺术的历史》，长沙：湖南科学技术出版社，第 203 页）

图 5-8　由理查德·诺曼·肖（Richard Norman Shaw）设计的伦敦老天鹅住宅建筑客厅室内设计（邵宏主编，颜勇、黄虹等编著《西方设计：一部为生活制作艺术的历史》，长沙：湖南科学技术出版社，第 203 页）

图 5-9　由沃尔特·克兰（Walter Crane）设计的伦敦莱顿勋爵别墅的室内一景（（美）大卫·瑞兹曼著，若澜达·昂，李昶译《现代设计史》，北京：中国人民大学出版社，2012 年，第 120 页）

图 5-10　由奥布里·比亚兹莱（Aubrey Beardsley）设计的《莎乐美》插图，运用东方艺术中的线条及装饰手法（邵宏主编，颜勇、黄虹等编著《西方设计：一部为生活制作艺术的历史》，长沙：湖南科学技术出版社，第 226 页）

图 5-11　安东尼·高蒂（Antoni Gaudi）设计的新艺术风格建筑，位于西班牙巴塞罗那（紫图大师图典编辑部《新艺术运动大师图典》，西安：陕西师范大学出版社，2003 年，第 55 页）

图 5-12　路易斯·多米尼科·蒙塔尼（Louis Domenico Montaner）的加泰罗尼亚音乐礼堂室内设计，位于巴塞罗那（紫图大师图典编辑部《新艺术运动大师图典》，西安：陕西师范大学出版社，2003 年，第 131 页）

图 5-13　维克多·霍塔（Victor Horta）设计的画室内景，位于布鲁塞尔的索尔维饭店（紫图大师图典编辑部《新艺术运动大师图典》，西安：陕西师范大学出版社，2003 年，第 87 页）

图 5-14　凡·德·维尔德（Henry Van De Velde）的优秀作品《天使的注视》（紫图大师

图版目录
List of Figures

图典编辑部《新艺术运动大师图典》,西安:陕西师范大学出版社,2003 年,第 205 页)

图 5-15　凡·德·维尔德(Henry Van De Velde)设计的书桌(紫图大师图典编辑部《新艺术运动大师图典》,西安:陕西师范大学出版社,2003 年,第 207 页)

图 5-16　查尔斯·伦尼·麦金托什(Charles Rennie Mackintosh)设计的椅子(紫图大师图典编辑部《新艺术运动大师图典》,西安:陕西师范大学出版社,2003 年,第 119 页)

图 5-17　查尔斯·伦尼·麦金托什(Charles Rennie Mackintosh)的室内餐厅设计风格(紫图大师图典编辑部《新艺术运动大师图典》,西安:陕西师范大学出版社,2003 年,第 123 页)

图 5-18　查尔斯·伦尼·麦金托什(Charles Rennie Mackintosh)设计的格拉斯哥艺术学院的校舍(紫图大师图典编辑部《新艺术运动大师图典》,西安:陕西师范大学出版社,2003 年,第 125 页)

图 5-19　路易斯·沙利文(Louis H. Sullivan)设计的信托银行大厦((英)丹·克鲁克香克主编,郝红尉等译《建筑之书:西方建筑史上的 150 座经典之作》,济南:山东画报出版社,2009 年,第 211 页)

第 6 章图:

图 6-1　14 岁留学欧美,毕业于美国哈佛大学建筑学院的设计师任天,现为中国美术学院建筑艺术学院教师

图 6-2　上横街村的三角亭改造之前的景象(实地拍摄)

图 6-3　上横街村的三角亭改造之后的景象之一(实地拍摄)

图 6-4　上横街村的三角亭改造之后的景象之二(实地拍摄)

图 6-5　上横街村的五芳亭改造之前的景象(实地拍摄)

图 6-6　上横街村的五芳亭改造之后的景象之一(实地拍摄)

图 6-7　上横街村的五芳亭改造之后的景象之二(实地拍摄)

图 6-8　上横街村的五芳亭改造之后的景象之三(实地拍摄)

图 6-9　上横街村改造之后的建筑立面坡檐之一(实地拍摄)

图 6-10　上横街村改造之后的建筑立面坡檐之二(实地拍摄)

图 6-11　上横街村改造之后的建筑立面坡檐之三(实地拍摄)

图 6-12　上横街村改造之后的建筑立面坡檐之四（实地拍摄）

图 6-13　上横街村的三间猪舍改造之前的实景（实地拍摄）

图 6-14　上横街村的三间猪舍改造之后的展厅（实地拍摄）

图 6-15　上横街村的三间猪舍改造之后的文化展厅，定期展出一些文化传统工艺项目

图 6-16　上横街村的碾米厂古建筑改造之前的实景（实地拍摄）

图 6-17　上横街村的碾米厂古建筑改造之前的室内景象

图 6-18　上横街村的三棵树碾米厂改造之后的文化空间之一（实地拍摄）

图 6-19　上横街村的三棵树碾米厂改造之后的文化空间之二（实地拍摄）

图 6-20　上横街村的三棵树碾米厂室内改造之后的文化空间之一

图 6-21　上横街村的三棵树碾米厂室内改造之后的文化空间之二（实地拍摄）

图 6-22　上横街村的三棵树碾米厂室内改造之后的文化空间之三（实地拍摄）

图 6-23　上横街村改造之后村口

图 6-24　上横街村改造之后的建筑细节设计之一

图 6-25　上横街村改造之后的建筑细节设计之二

图 6-26　上横街村改造的原创设计师、中国美院教师任天的乡村实践基地

第 7 章图：

图 7-1　住棚设计的文化由来，每年犹太人庆祝的三大节日之一

图 7-2　住棚节的传统文化，其间人们在棚内用餐、休息、聚会，欢庆节日

图 7-3　住棚设计外部的实地景观（原创设计师：中国美术学院教师任天）

图 7-4　住棚设计的原理图（原创设计师：中国美术学院教师任天）

图 7-5　住棚设计的总平面图（原创设计师：中国美术学院教师任天）

图 7-6　住棚设计的结构模型图之一（XYZ 坐标）（原创设计师：中国美术学院教师任天）

图 7-7　住棚设计的结构模型图之二（XYZ 坐标）（原创设计师：中国美术学院教师任天）

图 7-8　住棚设计的结构模型图之三（原创设计师：中国美术学院教师任天）

图 7-9　住棚设计的平面图（原创设计师：中国美术学院教师任天）

图 7-10　住棚设计的剖面图（原创设计师：中国美术学院教师任天）

图 7-11　住棚设计的结构展示图（原创设计师：中国美术学院教师任天）

图 7-12　住棚设计的序列细节展示图（原创设计师：中国美术学院教师任天）

图 7-13　住棚设计的正面和侧面的立面图（原创设计师：中国美术学院教师任天）

图版目录
List of Figures

图 7-14　住棚设计平面交通路径图（原创设计师：中国美术学院教师任天）

图 7-15　住棚设计建造过程图（实地拍摄）（原创设计师：中国美术学院教师任天）

图 7-16　住棚设计的创意构思步骤展示（原创设计师：中国美术学院教师任天）

图 7-17　住棚设计内部的实地景观（原创设计师：中国美术学院教师任天）

图 7-18　住棚建成的实地景观（原创设计师：中国美术学院教师任天）

图 7-19　《记忆山丘》的中国传统文化意象之一（原创设计师：中国美术学院教师任天）

图 7-20　《记忆山丘》的西方传统文化意象之一（原创设计师：中国美术学院教师任天）

图 7-21　《记忆山丘》的总平面设计图之一（原创设计师：中国美术学院教师任天）

图 7-22　《记忆山丘》的总平面设计图之二（原创设计师：中国美术学院教师任天）

图 7-23　《记忆山丘》的构造设计图（原创设计师：中国美术学院教师任天）

图 7-24　《记忆山丘》的构造细部图解（原创设计师：中国美术学院教师任天）

图 7-25　《记忆山丘》的局部构造模型（原创设计师：中国美术学院教师任天）

图 7-26　《记忆山丘》的整体构造模型（原创设计师：中国美术学院教师任天）

图 7-27　《记忆山丘》的建造现场（实地拍摄）（原创设计师：中国美术学院教师任天）

图 7-28　《记忆山丘》的建造过程（原创设计师：中国美术学院教师任天）

图 7-29　《记忆山丘》的现场实景（原创设计师：中国美术学院教师任天）

图 7-30　《记忆山丘》实景与现场观摩（原创设计师：中国美术学院教师任天）

图 7-31　《记忆山丘》实地观摩深受观众喜爱（原创设计师：中国美术学院教师任天）

图 7-32　《记忆山丘》作品细部特写（原创设计师：中国美术学院教师任天）

图 7-33　《记忆山丘》原创设计作品深受欢迎（原创设计师：中国美术学院教师任天）

第 8 章图：

图 8-1　工业产品设计师：朱昱宁（现为浙江工业大学设计艺术学院教师，浙江省工业设计技术创新服务平台设计部部长，杭州飞神工业设计有限公司设计总监）

图 8-2　《八宝吉祥扇》获得"浙江省工业设计大赛优秀奖"（原创设计师：朱昱宁（浙江工业大学设计艺术学院）、孙亚青（杭州王星记扇厂）、王冰）

图 8-3　《八宝吉祥扇》设计作品展示图（原创设计师：朱昱宁（浙江工业大学设计艺术学院）、孙亚青（杭州王星记扇厂）、王冰）

图 8-4　《八宝吉祥扇》作品设计图（原创设计师：朱昱宁（浙江工业大学设计艺术学院）、孙亚青（杭州王星记扇厂）、王冰）

图 8-5　毕业于浙江工业大学设计艺术学院、美国帕森斯设计学院工业设计专业的设计师郑昱

图 8-6　融合中国古典园林元素的作品《延绵·书桌设计》（原创设计师：郑昱）

图 8-7　《延绵·书桌设计》的细节展示图（原创设计师：郑昱）

图 8-8　《延绵·书桌设计》的材料种类与选材过程（材料实验者：郑昱）

图 8-9　《延绵·书桌设计》关于颜色变化的材料实验（材料实验者：郑昱）

图 8-10　《延绵·书桌设计》关于造型变化的材料实验（材料实验者：郑昱）

图 8-11　《延绵·书桌设计》关于透光性的材料实验（材料实验者：郑昱）

图 8-12　《延绵·书桌设计》关于触感的材料实验（材料实验者：郑昱）

图 8-13　《延绵·书桌设计》关于表面纹理的材料实验（材料实验者：郑昱）

图 8-14　《延绵·书桌设计》关于材料组合的实验（材料实验者：郑昱）

图 8-15　《延绵·书桌设计》关于海绵上色的材料实验（材料实验者：郑昱）

图 8-16　《延绵·书桌设计》设计思维与分析图（原创设计师：郑昱）

图 8-17　《延绵·书桌设计》体现设计方法的产品草图之一（原创设计师：郑昱）

图 8-18　《延绵·书桌设计》体现设计方法的产品草图之二（原创设计师：郑昱）

图 8-19　《延绵·书桌设计》体现设计方法的制作流程图（原创设计师：郑昱）

图 8-20　《延绵·书桌设计》设计展示图（原创设计师：郑昱）

图 8-21　《蒸汽加热餐具组合》设计展示图（原创设计师：郑昱）

图 8-22　融合传统中国饮食文化的作品《蒸汽加热餐具组合》设计展示图（原创设计师：郑昱）

图 8-23　中国传统饮食色香味俱全，品类齐全、五味调和、营养丰富

图 8-24　制作中国传统菜肴"蒸""煮"的烹饪方法伴随着中国传统饮食习惯

图 8-25　"蒸笼"起源于汉代，传承至今已有两千多年历史。本组设计从中国传统饮食器具中汲取设计灵感

图 8-26　融合传统中国饮食文化的作品《蒸汽加热餐具组合》探索实验图（原创设计师：郑昱）

图 8-27　《蒸汽加热餐具组合》设计作品的材料与造型实验图（原创设计师：郑昱）

图 8-28　《蒸汽加热餐具组合》设计流程与功能图（原创设计师：郑昱）

图 8-29　《蒸汽加热餐具组合》设计功能与创意展示图（原创设计师：郑昱）

图 8-30　集中国传统菜肴"蒸""煮"烹饪方法于一体的《蒸汽加热餐具组合》设计展示图（原创设计师：郑昱）

图 8-31　《蒸汽加热餐具组合》设计功能与步骤展示图（原创设计师：郑昱）

图 8-32　《蒸汽加热餐具组合》设计创意展示图（原创设计师：郑昱）

第9章图：

图 9-1　设计师：俞雪莱（毕业于中国美术学院获设计学学士学位，毕业于英国圣马丁艺术学院获设计学硕士学位。爱马仕（中国）资深橱窗设计师、如沐展示设计有限公司设计总监。2015年创立设计师品牌"SHIRLEY YU"）

图 9-2　视觉文化设计《甲骨文元素》（原创设计师：俞雪莱）

图 9-3　视觉文化设计《甲骨文与瑞典伏特加》（原创设计师：俞雪莱）

图 9-4　"中国结"元素的2010年世博会海报设计（原创设计师：俞雪莱）

图 9-5　"中国结"元素的海报设计（原创设计师：俞雪莱）

图 9-6　"中国泥塑"元素的《三只小猪与狼》跨界设计作品展示图之一（原创设计师：俞雪莱）

图 9-7　"中国泥塑"元素的《三只小猪与狼》跨界设计作品展示图之二（原创设计师：俞雪莱）

图 9-8　"中国泥塑"元素的《三只小猪与狼》跨界设计作品展示图之三（原创设计师：俞雪莱）

图 9-9　"中国泥塑"元素的《三只小猪与狼》跨界设计作品展示图之四（原创设计师：俞雪莱）

第10章图：

图 10-1　吕勤智教授（前排左五）《哈尔滨工业大学"老土木楼"建筑保护与再利用设计》项目荣获"中国建筑设计奖"银奖

图 10-2　《哈尔滨工业大学"老土木楼"建筑保护与再利用设计》项目展示图之一，百年建筑的历史照片（设计主持：吕勤智教授；设计团队：吕勤智教授科研团队；项目获奖："中国建筑设计奖"银奖，2013年）

图 10-3　《哈尔滨工业大学"老土木楼"建筑保护与再利用设计》项目展示图之二百年建筑改造之前的历史照片（设计主持：吕勤智教授；设计团队：吕勤智教授科研团队；

项目获奖："中国建筑设计奖"银奖，2013年）

图 10-4 《哈尔滨工业大学"老土木楼"建筑保护与再利用设计》项目展示图之三，百年建筑保护、更新与再利用设计方案（设计主持：吕勤智教授；设计团队：吕勤智教授科研团队；项目获奖："中国建筑设计奖"银奖，2013年）

图 10-5 《哈尔滨工业大学"老土木楼"建筑保护与再利用设计》项目展示图之四，百年建筑保护、更新与再利用设计方案（设计主持：吕勤智教授；设计团队：吕勤智教授科研团队；项目获奖："中国建筑设计奖"银奖，2013年）

图 10-6 《哈尔滨工业大学"老土木楼"建筑保护与再利用设计》项目展示图之五，百年建筑保护、更新与再利用设计方案（设计主持：吕勤智教授；设计团队：吕勤智教授科研团队；项目获奖："中国建筑设计奖"银奖，2013年）

图 10-7 《哈尔滨工业大学"老土木楼"建筑保护与再利用设计》项目展示图之六，百年建筑保护、更新与再利用设计方案（设计主持：吕勤智教授；设计团队：吕勤智教授科研团队；项目获奖："中国建筑设计奖"银奖，2013年）

第 1 章　导论
Chapter One: Introduction

[本章导读]

当代设计领域呈现跨学科的多元化发展，在追求科技创新的同时，我们也需要关注"文化"在设计成果中凸显的重要作用。中国五千年文明史积淀了丰厚的文化底蕴，领悟设计文化有益于开拓设计构思，从我国优秀的传统文化中汲取设计灵感，以现代设计思维统率传统文化艺术元素具有无限潜力，结合现代科学技术、新兴材料的创新机遇，提升与再现传统文化的精髓，从而达到古今设计的融合创新、和谐一致，产生举世瞩目的设计成果。

1.1 文化与文明不同语境的释词考辨

Textual Research on Culture and Civilization

“文化”与“文明”两词意思相近，平时经常混淆，甚至互用。在中国古典文献中，虽已出现“文化”一词，但是，它的使用并不频繁。相比而言，“文明”一词在古代文献中运用较为广泛，并且意思上更符合当代人们对于文化的定义，更确切地表达了当代人们所认同的文化内涵。

中国历史上对于文化的解释，基于中国古典文献的考辨，源自《子夏易传》:“观乎天文，以察时变；观乎人文，以化成天下。”(《子夏易传》一卷，春秋・卜商撰，清嘉庆四年（1799 年）平湖孙堂映雪堂刻本。或见清・通志堂经解本)。早在汉代“文化”一词已出现在刘向撰写的《说苑》第十五卷中：

凡武之兴为不服也，文化不改，然后加诛（图 1–1、图 1–2）。

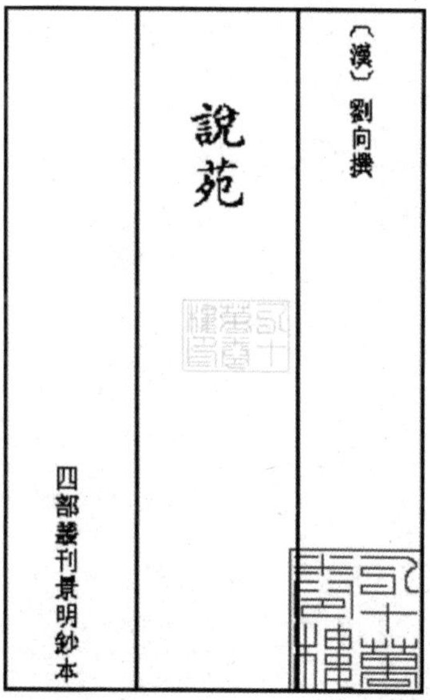

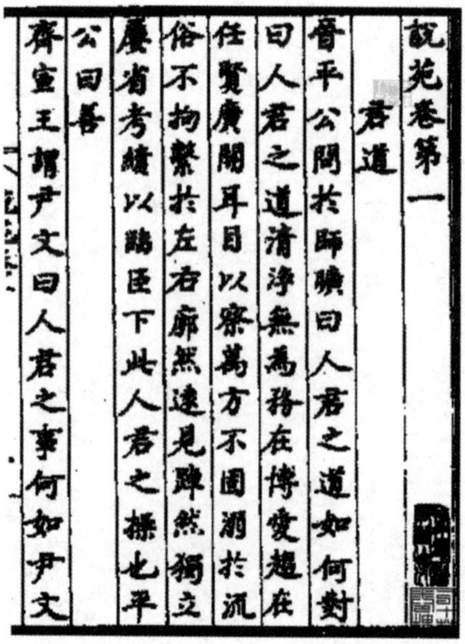
說苑卷第一
君道
晉平公問於師曠曰人君之道如何對
曰人君之道清淨無為務在博愛趨在
任賢廣開耳目以察萬方不固溺於流
俗不拘繫於左右廓然遠見踔然獨立
屢省考績以臨臣下此人君之操也平
公曰善
齊宣王謂尹文曰人君之事何如尹文

图 1–1　汉・刘向《说苑》封面（左）
（见：四部丛刊景明钞本。或见：清光绪元年（1875 年）湖北崇文书局刻本）

图 1–2　汉・刘向《说苑》正文页（右）
（见：四部丛刊景明钞本。或见：清光绪元年（1875 年）湖北崇文书局刻本）

近代赵尔巽撰写的《清史稿》也有关于“文化”的记载：

礼聘名儒为书院山长，其幕府亦极一时之选，江南文化遂比隆盛时（图 1–3、图 1–4）。

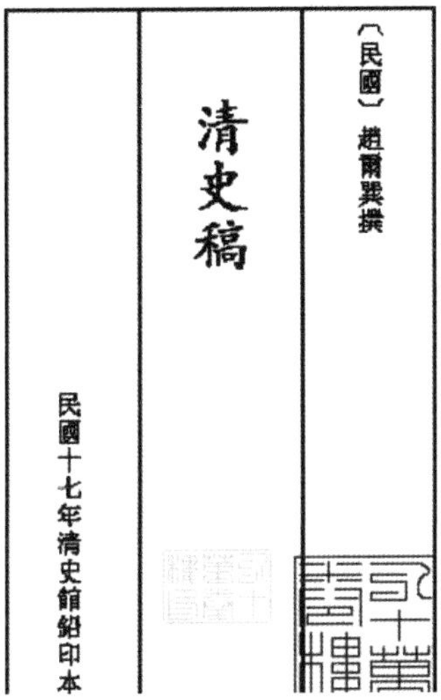

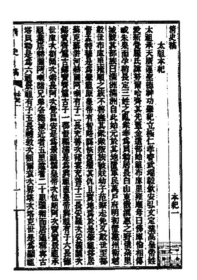

图 1–3　近代赵尔巽撰《清史稿》封面，第一百九十二卷（左）
（见：1928 年清史馆本。或见：1942 年联合书店刻本）

图 1–4　近代赵尔巽撰《清史稿》第一百九十二卷（右）
（见：1928 年清史馆本。或见：1942 年联合书店刻本）

“文化”一词在我国古代文献中尽管有所使用，但是，它出现的时间晚于“文明”一词，运用十分广泛，并且在意思上相当于现代人们所认同的文化内涵。周朝的《子夏易传》中有关于“文明”一词的记载：

天下文明终日乾乾（图 1–5、图 1–6）。

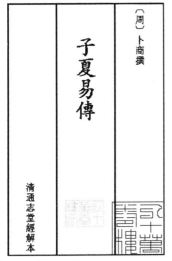

图 1–5　周朝卜商撰《子夏易传》卷一封面
（见：清通志堂经解本）

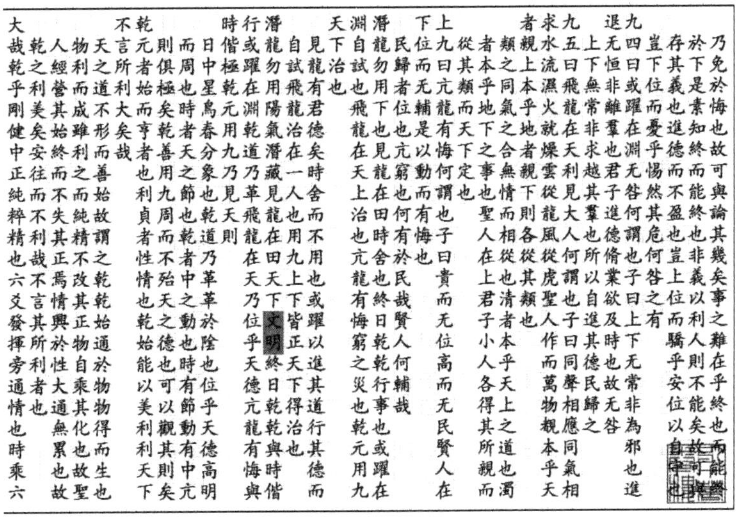
乃免於悔也故可與論其幾矣事之難在乎終也而能
於下是素知終而能終也非義以利人則不能矣故可
存其義也進德而不盈也豈上位而驕乎安位以自
豈下位而憂乎惕然其危何咎之有
九四曰或躍在淵无咎何謂也子曰上下无常非為邪也進
退无恒非離羣也君子進德脩業欲及時也故无咎
上下無常非求越其羣也所以自進其德民歸之
九五曰飛龍在天利見大人何謂也子曰同聲相應同氣相
求水流濕火就燥雲從龍風從虎聖人作而萬物覩本乎天
者親上本乎地者親下則各從其類也
類之同氣之合無情而相從也清者本乎天上之道也濁
者本乎地下之事也聖人在上君子小人各得其所親而
從其類而天下定也
上九曰亢龍有悔何謂也子曰貴而无位高而无民賢人在
下位而无輔是以動而有悔也
民歸者位也亢窮也何有於民哉賢人何輔哉
潛龍勿用下也見龍在田時舍也終日乾乾行事也或躍在
淵自試也飛龍在天上治也亢龍有悔窮之災也乾元用九
天下治也
見龍有君德矣時舍而不用也或躍以進其道行其德而
自試飛龍治在一人也用九上下皆正天下得治也
潛龍勿用陽氣潛藏見龍在田天下文明終日乾乾與時偕
行或躍在淵乾道乃革飛龍在天乃位乎天德亢龍有悔與
時偕極乾元用九乃見天則
日中星鳥春分象也乾道乃革革於陰也位乎天德高明
而周也時者天之節也乾者中之動也時有節動有中亢
則俱極矣乾善用九周而不殆天之德也可以觀其則矣
乾元者始而亨者也利貞者性情也乾始能以美利利天下
不言所利大矣哉
天之道不形而善始故謂之乾乾始通於物物得而生也
物利而成雖利之而純精不改其正物自乘其化也故聖
人經營其始終而不失其正焉情興於性大通無累也故
乾之利美矣安往而不利哉不言其所利者也
大哉乾乎剛健中正純粹精也六爻發揮旁通情也時乘六

图 1–6　周朝卜商撰《子夏易传》卷一
（见：清通志堂经解本）

宋代的《书经集传》中也使用了"文明"一词，诸如：中国文明之地，故曰华夏四时之夏（图1-7、图1-8）。

图1-7　宋代蔡沈撰《书经集传》封面
（见：清文渊阁四库全书本）

图1-8　宋代蔡沈撰《书经集传》
（见：清文渊阁四库全书本）

可见，在我国古代"文明"一词的含义实质上即指包罗万象的中国文化。"文化"在《辞源》中的释义是"文治和教化"。[1]"文化"一词通常英文理解为"culture"。英文culture的解析又分为三层含义：一是the ideas, customs, and art of a particular society；二是a particular civilization at a particular period；三是a developed understanding of the arts[2]。即：一是指特定的想法、风俗和艺术；二是指特定时期特定的文明；三是艺术的传承发展。从英文的释意来看，其本身包涵了"文化"与"文明"两层含义。

"文明"一词意为文采光明，文德辉耀；有文化的状态。[3]"文明"的英文解析是"civilization"。英文civilization的解析也分三层含义：一是the total culture and way of a particular people, nation, region, or period；二是a human society that has a complex cultural, political, and legal

---

[1] 广东、广西、湖南、河南辞源修订组、商务印书馆编辑部编《辞源》，商务印书馆出版，2009年9月第13次印刷，第1483页。

[2] Justin Crozier, Alice Grandison, Helen Hucker, Cormac McKeown Editors. (2008) Collins English Dictionary. Harper Collins Publishers. P185.

[3] 广东、广西、湖南、河南辞源修订组、商务印书馆编辑部编《辞源》，商务印书馆出版，2009年9月第13次印刷，第1485页。

organization；三是 intellectual, cultural, and moral refinement[①]。即：一是指特定人群、国家、地区和时间的整体文化；二是指具有文化、政治和法律组织共同融合的人类社会；三是完善的知识、文化与道德。从英文的释意来看，文明即是人类智慧、道德与文化的整体呈现。文明与文化的概念紧密相连，形影相随。

英文 culture 所指的含义也更深刻地体现在古典文献的"文明"一词之中。从释词可知，第一，中文和英文都将"文化"和"文明"两词的意义融会贯通起来；第二，英文 culture 的含义更侧重于我国古典文献中"文明"一词蕴含的意义；第三，当代"文化"一词的含义更为广博，无论中文，或是英文都表达了它与艺术相辅相成的内在关联，艺术与设计以丰富多彩的形式展现文化，成为文化的载体之一。

---

① Justin Crozier, Alice Grandison, Helen Hucker, Cormac McKeown Editors. (2008) Collins English Dictionary. Harper Collins Publishers. P135.

1.2 "设计"释词与古今考辨
Ancient and Modern Textual Research on Design

众所周知,"设计"一词由英文翻译而来,英文"design"一词在 Collins English Dictionary 的解析分为三层含义:一是 a sketch, plan, or preliminary drawing, 即草图、计划或初步图;二是 the arrangement of features of an artistic or decorative work, 即艺术或装饰物品的布局特征;三是 a finished artistic or decorative creation, 即艺术完成作品或者创造性的装饰;四是 to plan and make (something) artistically, 即有计划地使作品艺术化。

从石器时代开始,人类文化的进化与发展在可以称之为"造物文化"发展史,而"造物"一词本身就蕴涵了人类最初的设计概念。追溯我国古典文献中具有相近词意的表述,与现代设计(design)异曲同工、意义相近的概念是北宋李诫所著《营造法式》中的"法式"一词,专指方法、范式。称之为"造物""法式"或"设计"并不是最重要的,关键在于其实质是具有创新思想的行动。

在我国古代文献中,"设计"的概念一般分别为"设"和"计"两个独立的字,对于"设计"的考证,最早有文献记载的是许慎的《说文解字》一书中,对"设"与"计"两字分别作出阐释:"设,施陈也;计,会也,算也。"[①] "设"与"计"均属会意字,具有核算、测量、策略、打算或姓氏等意思。这两个字与近、现代人们所理解的"设计学"含义相差甚远。"设计"作为一个完整的词汇曾用于《三国志·魏书·三少帝纪》一书,意思是"筹划计策"。然而,这个词汇在古代的用法和意思与现代人们所理解的"设计学"含义截然不同。直至20世纪初期,《辞海》巨著首次问世,才形成了较为完善契合的解析。"设计"一词在《辞海》中的释义是:根据一定的目的要求,预先制定方案、图样等活动。我们将现代设计理解为"在正式做某项工作之前,根据一定的目的要求,预先制定方法、

---

① 许慎《说文解字》,中华书局,1978年版,第53页。

图样等"①从学科定义上来看，设计，是运用美学原理，按照产品的功能要求而策划、制定的方案、图样。它是一门将物质文明与精神文明、科学技术与文化艺术高度结合的应用学科。可见，"设计"在我国是诞生于20世纪的一个新词。

追溯"设计"一词的由来，陈之佛在《图案构成法》（1937年版）一书中写道："图案在英语中叫design。这design的译意是'设计'或'意匠'，所谓'图案者'，是日本人的译意，现在中国也普遍地通用了。"卢世主也提出："中国早期把英语Design译为'图案'，这是沿用日本的译法。称图案为'图的考察'，所以图案也叫做'考察画'。"②当代设计领域日新月异，我国高校艺术设计专业普遍以"引进西方，中西合璧"为发展趋向，教学内容不仅有外来设计表述之类"术语"的英汉互译，还常常出现大量反映东西方文化艺术差异的原创设计概念。

实质上，设计同仁们不要误以为我国的"设计学"是一门诞生于20世纪80年代的新兴学科。"设计学"的造物思想、营造理论与实践运用在中国历代文献中均有记载。只是古今释词仅仅在称谓上不同而已，自石器时代以来，中华文化的演进与发展在相当长的时期内实际上孕育了一部营造设计、造物文化的历史。我国古代能工巧匠的创新成就举世闻名，可以追溯到春秋战国时期的制车、礼器等设计，甚至更为久远，其蕴涵的设计观念与方法亦在历代建筑、园林、家具、器物、用品中施展沿用。

因为"设计"这个新词，我们往往忽略了"设计"与传统文化的联系，我们不能因为"设计"这个新词而隔断了它与中国五千年文明史之根连。近年来一些设计作品单单追求时尚效应，过多地关注当代设计求新、求异的呈现形式，却失去了许许多多向古代传统文化取经的机会。然而，西方的设计大师们早已发现"文化"的凝聚力与创新力，"文化"成为一种新能源，带给现代设计领域无限的活力，植入文化内涵的设计，不仅把握设计作品之"形"，更在于体现其"质"。在当代国际设计舞台上，西方的古代文明不断地再现于新兴设计之中，诸如：由詹姆斯·斯特林③设计的斯图加特的新国立美术馆建筑，将"传统古希腊柱式"巧妙地引用到美术馆入口设计中，与现代建筑风格相融合，达到了古今建筑形制的和谐一致。从建筑设计、室内装饰、文学艺术、多媒体影像等各个领域、各个层面，结合现代科学技术、新兴材料的创新机遇，不断地进行翻新与提升，再现传统文化的新兴设计。

我们从中国古代文明的视角来考证"设计"一词，必然发现中华文化一脉相承、源远流长，我国古代的造物思想、营造创新更是一脉相传、世代相承。笔者认为，设计文化由历史积淀而形成，它是一种承载创造灵感、凝聚传统文化、力求传承出新的创意文化。

---

① 中国社会科学院语言研究所《现代汉语词典》，商务印书馆，2005年，第1203页。
② 卢世主《从图案到设计：20世纪中国设计艺术史研究》，江西人民出版社，2011年，第179页。
③ 詹姆斯·斯特林（James Stirling, 1926—1992）是英国著名建筑设计师，也是1981年建筑普立兹奖的获奖者。

第2章 中国古典设计文献选读

Chapter Two: Selected Readings in Classical Chinese Design Text

[本章导读]

如果留心观察，我们会发现当代很多产品设计、宣传品设计、展示设计、建筑以及室内设计普遍追求新异，单纯地去追求一些新奇效果往往会造成设计"形式化"，而无法满足人们高层次的审美需求。这些现状对当代及未来设计提出了值得思考的新问题。如何来改善这种现象？我们需要弥补设计作品文化内涵的不足，以中国传统文化作为基石，积极探索与开创具有高层次文化品位的设计作品。五千年悠久的中华文明史告诉我们，不是我们的传统文化底蕴不足，更不是传统文化元素不够，而是目前设计者们采集、挖掘、开拓与运用得太少。

本章从古代文献中精选八部留存至今，体现我国古代文化与设计思想的典籍节选并加以注释及解析。从历代皇宫贵族到文人雅士生活中必备的居室布置、家具陈设、器物用品、文房四宝，往往反映了最为真实的文化印迹、最为朴素的造物思想。古时能工巧匠们设计精美的皇室御用珍品是中华文明的瑰宝，也是古代文明的印证。清代人查继佐在其撰写的《徐光启传》中云："观时深而验物切……"，蕴含了富有远见的设计思想。设计与文化是一个有机整体，犹如"大树与土壤"，文化是设计的根基，也是设计之树四季常青、枝繁叶茂的源泉。我们将设计与文化相融互训，以现代的方式传承民族传统文化，并运用中国传统文化创生出我们这个时代物质与精神结合的原创设计作品，犹如生命中的美好基因，成为未来中国设计的原动力。

2.1 宋·郭熙《林泉高致》(节选)

Lin Quan Gao Zhi. Author Guo Xi. Song Dynasty (Selected Sections)

"世之笃论：谓山水有可行者、有可望者、有可游者、有可居者，画凡至此，皆入妙品。但可行、可望、不如可居、可游之为得。何者？观今山川，地占数百里，可游、可居之处，十无三四，而必取可居、可游之品。君子之所以渴林泉者，正谓慕此处故也（图2-1、图2-2）。"

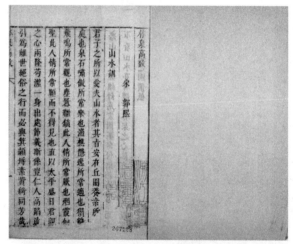

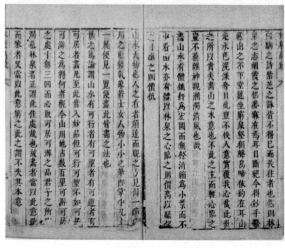

图2-1 宋·郭熙《林泉高致》，明代读书坊刻本，封面（上）

图2-2 宋·郭熙《林泉高致》，明代读书坊刻本，第1-2页（下）

世人有一个确切顺理的看法：说山水，有些可以行走其间，有些可以远远观望，有些可以尽情畅游，有些则适宜安居在此地，山水画凡是能够达到这种境界，都属于妙品佳作。但是，可以行走其间，可以远远观望，不如可以尽情畅游、可以适宜安居更合人意。什么原因呢？看看现在的山川，占地几百里，可以尽情畅游、可以适宜安居的地方还不到十分之三四，但我们一定要去寻找那可以尽情畅游、适宜安居的最佳之地所在。君子之所以渴望归隐林泉，正是非常向往思慕这样的好去处。

此段的关键之处"但可行、可望，不如可居、可游之为得"，此句指明了我国古代园林的建造崇尚"自然主义"，往往选址在可以恣意畅游、适宜栖居的场所，在园林中，人们既可驻足品玩，又能够步移景异地欣赏美景，让归隐林泉的安居者流连忘返、美不胜收。作为一部以描绘山水画为主体的画论著作，其中体现了作者的中国古典园林设计思想，他注重古典文化气息，将人文气质再现于山水表现之中，他讲究"身心合一"、情景交融、诗情画意的空间想象与意境表达，使诗词为造园之理，绘画为造园之图。

There has been a view in the world of landscapes that is both accurate and sensible. It holds that some landscapes are meant for people to wander in, some are suitable for distant appreciation, some make excellent destinations for sightseeing, and some are good for people to reside in. All landscape paintings that belong to one of the four listed areas are excellent works of art. However, those suitable for people to wander in or for distant appreciation are not as good as those best for sightseeing or for people to reside in. What's the reason then? Look at the mountains and rivers today. They might occupy an area of hundreds of miles, quite suitable for sightseeing, but only 3/10 to 4/10 of the area is inhabitable. People are determined to find the best location where they can both enjoy sightseeing and build up a residence. The reason why gentlemen desire to retreat into seclusion is simply that they yearn for such an ideal destination.

The key point in the paragraph above is that "Landscapes suitable for people to wander in or for distant appreciation are not as good as those best for sightseeing or for people to reside in." It points to the central idea of cherishing naturalism in landscape architectures in ancient China so that people could wander wherever their instinct took them while residing properly. In the garden, the retired scholar could both stop for a while to appreciate the surrounding beauty and enjoy a different scene at a different step. As a major work featuring landscape paintings, Guo Xi's *Lin Quan Gao Zhi* presents the author's thoughts on the designing of the Chinese classical gardens. He lays emphasis on classic cultural flavor while embedding humanitarian temperament in landscape expressions. What's more, he also focuses on the harmony between body and mind, on the blend of sentiments

郭熙的《林泉高致》不仅是一部宋代山水画品评标准的画史论著，也是中国古典园林造园构架的一种审美标准。我国历代的文人雅士既是山水画的创作者、欣赏者，又是山水园林建筑的设计者、居住者。郭熙在画论中阐释的"静虚""如涉其中"和"三远"的画境与古典园林建筑所营造的意境是一致的，所追求的设计理念是一致的，所展现的文化根基也是一脉相承的。尤其在宋代，人们选址建造园林是以山水画作为造园艺术的参照标准。英国皇家建筑师威廉·钱伯斯（1723-1796）曾说："中国造园匠不是花儿匠，而是画家和哲学家。"他为伦敦郊区的丘园（Kew Garden）建造"中国塔"时同样参照了中国山水画中的造园艺术。

and scenery, and on the merge of poetic space imagination with environment expression. Thus he treats poems as the basis for the garden architecture and paintings as the drawing sketch for the landscape design.

Guo Xi's book titled *Lin Quan Gao Zhi* 《林泉高致》 is not just a book on the history of paintings discussing the standards of judging Song Dynasty landscape paintings, but it sets the aesthetic standards for building and judging classical Chinese gardens. Refined scholars and literati in the past dynasties in China were not only the creators and appreciators of landscape paintings, but they were also the designers and dwellers of landscape gardens and buildings. Guo Xi's theories of painting such as silence and void, immersive paintings, and three kinds of farness of overlooking, profound, and plane observation in Chinese paintings are all in line with the ambience created by classical garden buildings, with the designing concepts sought by those buildings, and with the cultural roots expressed by them. Especially in the Song Dynasty, people chose a site for building a private garden by referring to landscape paintings as standards to go by. William Chambers (1723-1796), the British royal architect, said that instead of the gardener, the landscape architect masters in China were more than the painter and philosopher. Chambers also built the Chinese Tower in the Kew Garden in the suburbs of London by drawing inspirations from the garden-designing techniques shown in Chinese landscape paintings.

2.2 元·《大元毡罽工物记·御用》(节选)

Da Yuan Zhan Ji Gong Wu Ji·Yu Yong. Yuan Dynasty (Selected Sections)

"至治三年十二月五日,留守伯胜阿鲁泽沙传旨,北平王影堂内核桃木碗碟、象牙匙、筋楠木桌及诸物,依世祖皇帝影堂制,从新为之,计料绘图成造(图2-3、图2-4)。"

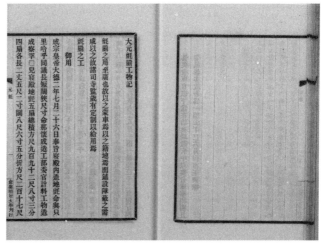

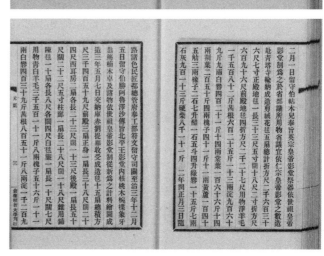

图2-3 元·佚名《大元毡罽工物记·御用》,1916年,上海仓圣明智大学铅印本,封面(上)

图2-4 元·佚名《大元毡罽工物记·御用》,1916年,上海仓圣明智大学铅印本,第3-4页(下)

在 1323 年（元代至治三年）农历十二月五日，留守（官位）伯胜阿鲁泽沙传达旨意：北京王影堂里所使用的核桃木材质的碗和碟子、用象牙制成的勺子、楠木材质的桌子以及其他物件，按照世祖皇帝供奉神佛的影堂制造，制作时唯求新意，先计算材料的用量、绘制成图后再制造出来。

此段话强调了我国古代"从新为之"的造物思想。一方面中国古代通过器物、物件来体现长幼尊卑的地位，按照钦定的法度进行设计和生产，使每一件产品的款式、造型、纹样、色彩、工艺、材质都具有明显的设计识别性；另一方面也说明了古代工匠已具有制造不同规格的工艺品的高超技能，在制造工艺、器物式样、物件材质等方面力求推陈出新、与众不同的形质，反映唯物求新、勇于创新的设计思想。

《大元毡罽工物记》的作者不详。此书全面而详细地记载了元代纺织品的手工业生产系统，皇室采用的毡毯名目、样式以及生产毡毯所用的各种纺织材料与特点阐释，论及元代的社会文化、工艺技术与设计观念等各方面情况，对毡罽工艺记载翔实，具有史料价值。

Bo Shen A Lu Ze Sha, the Official positions of Leftover conveyed decree on 5th December 1323 in Yuan dynasty as follows: the furniture and articles in Beijing Wang's Buddhism hall were all manufactured according to the standard of the Hall of Images for the Emperor worship to the Buddha, the walnut bowls and plates, spoons made of ivory material, as well as nanmu wood tables and articles pursuing for the innovation during manufacturing were produced after calculating the material amount and planning drawings.

This statement emphasizes "Pursuing Novelty in Creating Things", the ancient Chinese thought on creating things. On one hand, positions of high and low, gentle and simple were embodied through using implements, utensils and articles in ancient China. All design and production was according to laws made by imperial order so each piece of product must have obvious design identity in its style, model, vein, color, technology and material aspects. On the other hand, it is proved that the ancient craftsmen has possessed highly skills for producing handicraft articles with different specifications and pursued for new styles and unique forms during manufacturing technology, utensil styles and object materials, which could reflect a thought of innovation and creation.

The author of the book titled *Da Yuan Zhan Ji Gong Wu Ji* (*Article Records of Worker of Carpet and Blanket in Yuan Dynasty*) is unknown. This book fully details the production system of textile handcraft industry in Yuan dynasty including the carpet types and styles used for imperial household as well as various textile materials and features of carpet production, and describes the social culture, technology and design concepts of Yuan dynasty and so on, contributing to the historical value of carpets and blankets technology records.

2.3 明·宋应星《天工开物》（节选）
Tian Gong Kai Wu. Author Song Yingxing. Ming Dynasty（Selected Sections）

"天覆地载，物数号万，而事亦因之，曲成而不遗。岂人力也哉？事物而既万，必待口授成而后识之，其与几何？万事万物之中，其无益生人与有益者，各载其半（图2-5～图2-9）。"

天地养育包容万物，世上的物类数以万计，因此，事物的纷繁复杂便由此衍生，天地以各种方式成就万物而无遗漏。这难道是人力可以相提并论的吗？事物既然有上万种那么多，必须等到别人口头传授和自己亲眼所见，然后才了解，那样能够知道多少呢？世间万事万物之中，对人们没有益处与有益处的状况，各占一半。

这段话比喻天地广阔，恩泽深厚，万物生长，妙不可言。"曲成而不遗"寓意人们要顺应自然规律、顺应事物的变化，大自然蕴藏其自身规律。但是，人们不是唯独等待大自然的恩惠，而是将"天工"与"人

The universe raises and holds everything. There are tens of thousands categories in the world. The diversity and complexity of things are thus derived. The universe accomplishes all things without any omission. Could manpower be mentioned in the same breath? There are thousands of things in the world. How much could one know only through word of mouth and witnessing? Among all things on earth, those that are advantageous to people and those that are disadvantageous to people account for one half, respectively.

This is a metaphor as follows: the vast heaven and earth, great blessings and the growth of all living things are ingenious beyond description. When melodies come to an end without left means and conditions are ripe, success will come. People should conform to natural laws and the natural changes of things. The nature has its own laws. However,

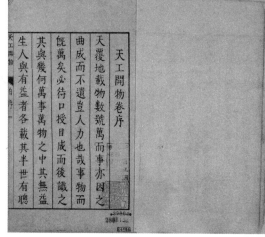
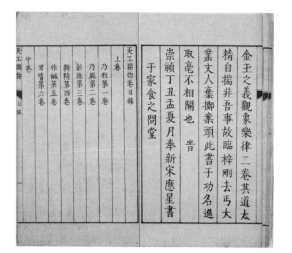
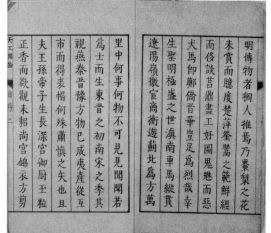
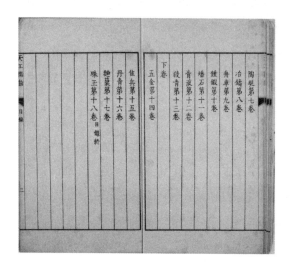

图2-5 明·宋应星《天工开物》，1930年，上海华通书局影印本，封面（左上）

图2-6 明·宋应星《天工开物》，1930年，上海华通书局影印本，目录第1页（左中）

图2-7 明·宋应星《天工开物》，1930年，上海华通书局影印本，目录第2页（左下）

图2-8 明·宋应星《天工开物》，1930年，上海华通书局影印本，自序第1页（右上）

图2-9 明·宋应星《天工开物》，1930年，上海华通书局影印本，自序第2页（右下）

力"相互契合，造就多元化的大千世界，其中不仅蕴涵人与自然和谐的"天人合一"思想，更是古人造物创新思想之源泉。

《天工开物》是明代宋应星撰写的一部记载我国古代农业和手工业生产技术、工艺装备、造物方法的综合性著作。"天工"意为自然力和人工的合力；"开"具有通畅、开拓、发展、建立之意；追求能工巧匠与自然条件的相互协调、相互生发，从而造就"物"。可见，中国传统造物文化中普遍具有遵循本源、顺应自然、物善其用、唯物求实的思想内涵。此书还论及陶瓷、车船、金属、农具等的造物方法，图文并茂、鉴古至今，被西方学者誉为"17世纪中国工艺百科全书"。英国学者李约瑟把宋应星誉为"中国的狄德罗"。①

Instead of waiting for the nature's blessings, manpower and work of nature should be integrated with each other to create a diversified big world. Here not only shows the harmony with nature - the idea of Unity of Man and Nature's but also the source of the ancients' natural innovation ideology.

The Book titled *Tian Gong Kai Wu* is written by Song Yingxing in the Ming Dynasty. It is a comprehensive works and records ancient agriculture, product techniques of handcrafts, Craft equipment, and methods of creation. The phrase "*tiangong*" implies the joint effects of natural force and human force and "*kai*" means expediting, expanding, developing and building. The "*Wu*" can be created by pursuing the mutual coordination and development between skillful craftsman and natural conditions. It is therefore observed that the Chinese traditional creation culture reflects the ideological implication of complying with source, conforming to nature, making the best of everything and taking a practically materialistic attitude. Moreover, the creation methods for boats and cars, metal, farm implements etc. are also illustrated in this book. Due to the combination of vivid pictures and descriptions as well as the references based on ancient experience, it was praised as "the technique encyclopedia of China in the 17th century" by Western scholars. The book's author Song Yingxing got praise as the "Denis Diderot of China" by the British scholar Joseph Li.

---

① 狄德罗（Denis Diderot,1713—1784）：18世纪法国启蒙思想的领袖，主编《百科全书》。

2.4 明·计成《园冶·装折》（节选）
Yuan Ye·Zhuang Zhe. Author Ji Cheng. Ming Dynasty (Selected Sections)

"凡造作难于装修，惟园屋异乎家宅，曲折有条，端方非额，如端方中须寻曲折，到曲折处还定端方，相间得宜，错综为妙（图2-10～图2-12）。"

图2-10 中国园林建筑的牛腿
（手绘创作者：陈炜副教授，浙江工业大学设计艺术学院）

第 2 章　中国古典设计文献选读

Chapter Two: Selected Readings in Classical Chinese Design Text

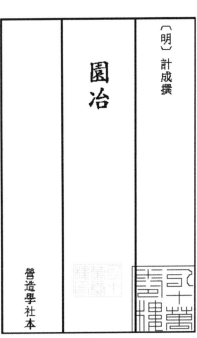

图 2-12　明·计成《园冶》三卷，1932 年，中国营造学社铅印本，封面

图 2-11　中国园林建筑的牛腿
（手绘创作者：陈炜副教授，浙江工业大学设计艺术学院）

凡是建筑，都难在装饰上，园林的装饰与住宅的装饰是不同的。曲折中要有条理，不必完全端端正正的。比如端正之中必须找到曲折的地方，曲折的地方也必定能找到有条理的端正之处。两者相互渗透、纵横交错、彼此适宜，才是最好的构筑。

园林建筑与住宅居室的装修手法不同，它们蕴藏的审美文化也各不相同。园林建筑讲究统一中求变化，变化中不失统一的布局；园林建筑善于运用复杂的曲线、不规则的外形，汲

All buildings are hard to be decorated. Garden architecture decoration is different from residence decoration. Order should be incorporated with intricacy, and it is unnecessary to be exactly regular and upright. In order words, intricacy must be found in order, while order can also be perceived in intricacy. It will be optimal if both of them can be mutually penetrated, interweaved, and suitable for each other.

Landscape architecture and residential rooms have different decoration means, containing different cultural aesthetic standards. Landscape architecture emphasizes on the overall arrangement of changes in unity and unity in change. Landscape architecture well utilizes complicated curves and irregular

取古代书画中的文化元素，并且营造与室外造景相互融合、和谐统一、具有中国诗画意境的景观。

明代不仅兴建了许多传承至今的著名园林，还成就多部对研究我国古代园屋设计具有史料价值的造园著作，计成撰写的《园冶》就是其中之一。计成的创造性设计造诣很深，他不墨守成规、不生搬硬套别人的一些既定法则。自古以来，造园与建造领域追求勇于创新，创生自己别具一格的新设计；计成之成就，不仅仅在于设计，他还熟练掌握造景、构园的方法，在实践中得心应手地指挥工匠。他不愧是古代无人可比的具有艺术文化涵养与鉴赏能力的园林设计大师。

appearances, absorbs the cultural essence of ancient paintings as well as create the Chinese-style idyllic and artistic landscape which could reflect the harmony and unity between indoor decoration and outdoor environment through integration.

In Ming Dynasty, on one hand, many well-known heritage gardens were built being bequeathed now. On the other hand, some historical value monographs were achieved on studying of Chinese ancient garden design during the ages. The book titled Yuan Ye written by the author Ji Cheng is one of valuable monographs. Ji Cheng has great attainment and courage in creative design to innovate his new designs. He does not stick to convention or simply follow the established rules. Since ancient times, he has been devoted himself in innovation in the gardening and building area, creating a unique style of new designs. In addition to Ji Cheng's achievements in design, he also makes himself master of landscaping and constructional techniques, expertly commanding the craftsmen with great facility. With no doubt, he is an unparalleled master of landscape architecture well cultured in Arts with great connoisseurship.

2.5 清·李斗《工段营造录》(节选)

Gong Duan Ying Zao Lu. Author Li Dou. Qing Dynasty (Selected Sections)

斗科

"斗科做法,有平身科、柱头科、角科及内里棋盘板上安装品字科、隔架科之分。算斗科上升斗栱翘诸件长短高厚尺寸,以平身科迎面安翘昂斗口宽尺寸为度,有头等寸至十一等寸之别。头等六寸,以下降一等减五分(图2-13~图2-18)。"

斗栱的制作方法,有平身斗栱、柱式斗栱、角形斗栱以及内部棋盘板上安装的有"品"字形斗栱、隔架斗栱的区分。计算斗栱的上部由坐斗向外伸出的翘栱的各种构件的长短、高低、厚薄的尺寸,以平身斗栱,正面的由坐斗向外伸出的翘栱的直径宽度为尺度,分首等寸至十一等寸的区别。首等为六寸,以下依次降低一等减去五分。

Making methods of bucket arches have all kinds of varieties such as flat level bucket arches, column type bucket arches, shape of horn (angle) bucket arches, and the indoor chess board shape of "品 pin" character pattern bucket arches and rack bucket arches. In order to calculate the length, height, and thickness of the various members of warping arches stretching from seating buckets to the outside of upper brackets, it is taken flat brackets and the diameter breadth of outright warping arches stretching from seating buckets to the outside as a measurement which would be divided into chief inches to eleven inches.

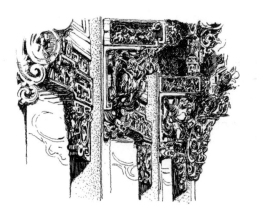

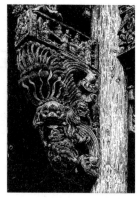

图 2-13 古建筑的斗栱
（手绘作者：陈炜副教授，浙江工业大学设计艺术学院）（左）

图 2-14 古建筑的牛腿
（与斗栱相连部分）
（手绘作者：陈炜副教授，浙江工业大学设计艺术学院）（右）

中國營造學社捐
儀徵李斗遺著
工段營造錄
朱啟鈐題

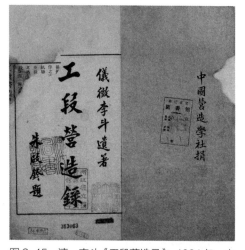

图 2-15 清 · 李斗《工段营造录》，1931 年，中国营造学社铅印本，封面

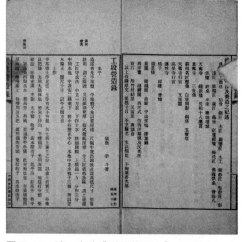

图 2-16 清 · 李斗《工段营造录》，1931 年，中国营造学社铅印本，正文页

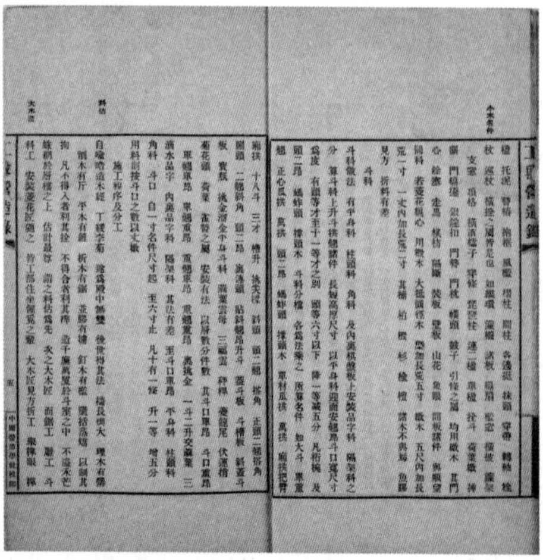

图 2-17 清 · 李斗《工段营造录》，1931 年，中国营造学社铅印本，斗栱，正文页

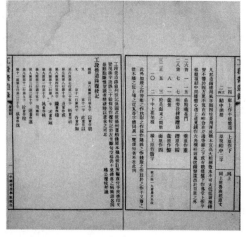

图 2-18 清 · 李斗《工段营造录》，1931 年，中国营造学社铅印本，封底版权页

第 2 章 中国古典设计文献选读
Chapter Two: Selected Readings in Classical Chinese Design Text

《工段营造录》成书于清乾隆六十年（1795年），收入《扬州画舫录》第十七卷，作者为清代李斗。李斗，清乾隆年间人，字艾塘，江苏仪征人。自幼好学，遍游大江南北，精通诗词、音律、数学。传承至今的论著主要有《扬州画舫录》《扬州名胜录》《艾塘曲录》等。

此书共分49章，记述了清代营造标准、建造方法与工程法则，特别是扬州地区的工段营造法则，可见清代颁布《工程做法》的要义。总体来看，这是关于营造技术及其工程规则方面的详细记述。具体分为：土作、大木作、折料法则、斗科、施工程序及分工、木材比重、搭材作、瓦作、砖作、琉璃瓦料、石作、裹角法、顶、装修作，并且论述了桥梁、亭、榭、楼、斋等形制以及琉璃影壁、木顶格、铜铁作、油漆作、画作、灯彩、室内陈设器玩、宫室释名等规格方法。从其瓦作、砖作、琉璃瓦料、装修作、画作、灯彩、室内陈设器玩多个部分，反映了当时的建筑文化印记，并对建筑制度提出一些规范法则。

The chief refers the six inches, and the followings are degraded by one and reduced by five points.

The book titled *Gong Duan Ying Zao Lu* was written by the author Li Dou Qing Dynasty in 1795 (Qianlong 60 years in the Qing Dynasty). It was collected in the seventeenth volume of the *Selections of Yang Zhou Gail-Painted Pleasure-Boat*. Li Dou styled himself Ai Tang came from Yi Zheng in Jiangsu Province during Qianlong period in Qing Dynasty. The author, travelling across the country, was keen on learning since childhood and proficient in poetry, temperament and mathematics. His monographies have been inherited so far are mainly the *Selections of Yang Zhou Gail-Painted Pleasure-Boat*, *Selections of Yang Zhou Scenic Beauty*, and *Selections of Ai Tang Melody*.

This book is divided into forty-nine chapters and includes standards of construction, methods of construction, and rules of engineering in the Qing Dynasty, especially rules of section construction in Yangzhou area. It is thus evident to show the key points and meanings of *Method of Engineering Work* Promulgated in the Qing Dynasty. In general, here are detailed reviews of construction technologies and engineering rules. They include earth tectonics, carpentry work, folding rule, bucket arch, process and work distribution of construction, specific weight of wood, scaffolding, tile-roofing, brick work, glazed tile, stone work, angle wrapping, top work and decoration work. The book also put forward arguments on the shape of bridge, pavilions, terrace, building and temple as well as the standard method of glazed screen wall, wooden roof, copper and iron work, painting work, drawing work, colored lanterns, indoor furnishes and displays and the name making and explaining of palaces. Imprinting of architectural culture could be mirrored from multiple parts such as the tile-roofing, brick work, material of glazed tile, decoration work, painting work, colored lanterns, indoor furnishes and displays.

2.6 清·寂园叟《匋雅》（节选）

Tao Ya. Author Ji Yuansou. Qing Dynasty
(Selected Sections)

"园蔬逾珍馐，瓦缶胜金玉，所谓别致者也（图2-19、图2-20）。"

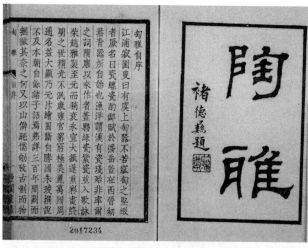

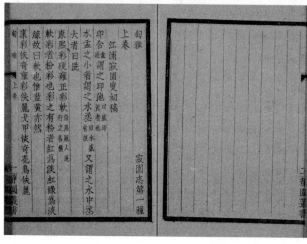

图2-19　清·寂园叟《匋雅》，1918年，静园刻本，自序页（上）

图2-20　清·寂园叟《匋雅》，1918年，静园刻本，第1页（下）

家园里自己种植的蔬菜，胜过珍奇贵重的食物。用泥土烧制的器具胜过用金子白玉制造的器皿。这就是与众不同、新奇有致的区别所在。

园中的果蔬未必不如山珍海味，各有效用、见仁见智，体现一种淳朴求真的生活态度。精雕细琢的玉器、流光溢彩的金银宝物，凭借晶莹华丽的外表而彰显其观赏价值，但陶土烧制而成的器物又何尝不如金玉。相比而言，陶器纯净质朴、拙中见雅，别有一番品鉴之味。倘若在室内摆设几件彩陶罐，恰好呈现"别具一格"的室内陈设艺术之美。

寂园叟撰写的《匋雅》是一部清代记述我国古陶起源、历代名窑、匠作、装饰、瓷器样式及其制作工艺的名著，也是一部反映古代瓷器文明的史书。对了解中国陶瓷起源、陶瓷文化以及制陶工艺发展史都具有很好的史料价值。

The home vegetables planted by own ever surpass those rare and precious foods. The utensils fired by clay even surpass those made from gold or white jade. This is the differences of uniqueness and novelty.

The vegetable and fruit in the garden has various effects based on various needs and is definitely not worse than delicacies of mountain and sea, which would shou a simple and validity attitude of life. The delicate jade and exquisite treasures show its ornamental value through their magnificent appearances. However, the implements made by clay are not any worse than gold and jade. By contrast, earthenware is pure, plain, defectively elegant and thus unique in appreciation. If a few pieces of pottery jars are decorated indoors, they could just present unique beauty of interior furnishing art.

This book titled *Tao Ya*. Written by the author Ji Yuansou is not only a masterpiece that records the origin of Chinese antique pottery, the famous kilns of past ages, the construction of craftsman, the decorations, the styles of porcelain and the workmanship in the Qing Dynasty, but also a mirror of historical records for ancient porcelains. It provides valuable historical data about the origin of ceramics, ceramic culture and the development of pottery techniques.

2.7 清·李渔《闲情偶寄·器玩部》（节选）
Xian Qing Ou Ji. Author Li Yu. Qing Dynasty
(Selected Sections)

"凡制茗壶，其嘴务直，购者亦然，一曲便可忧，再曲便称弃物矣。盖贮茶之物，与贮酒不同，酒无渣滓一斟即出，其嘴之曲直可以不论；茶则有体之物也，星星之火，入水即成大片，斟泻之时，纤毫入嘴，则寒而不流（图 2–21~图 2–26）。"

凡是制作茶壶，壶嘴一定要直，选购的时候也要按照这个标准，壶嘴一旦弯曲（它的）功能就令人担忧，如有两重弯曲简直就成为废弃物了。因为储存茶的器物和储存酒的器皿不一样，酒没有残余物，一倒就出来，因此酒壶嘴部的曲直造型可以不讲究；然而，茶是有形体的食物，即使是星星点点的小片茶叶，放入水中就变成大片（的茶叶），倒茶的时候，容易进入壶嘴，导致壶嘴被塞不易疏通。

The spout must be straight when making teapot. I must do according to this standard when purchasing on selection. If the spout is bended, its functions will be affected, and If there two bended areas, it will become useless. There utensil for tea is different from that for wine, there is no residual in wine, and once you pour it, it will come out, because there is no particular requirement for the spout shape of wine pot; however, tea is solid food, even small tea leaves will become bigger in water. When you pour the tea, bigger tea leaves tend to enter the spout place and block the spout.

第 2 章　中国古典设计文献选读
Chapter Two: Selected Readings in Classical Chinese Design Text

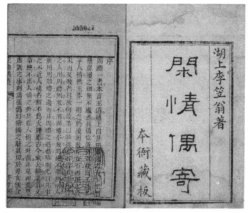

图 2-21　清·李渔《闲情偶寄·器玩部》，清康熙十年（1671）刻本，封面

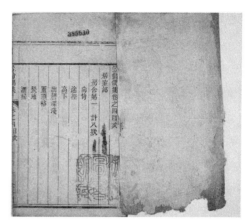

图 2-22　清·李渔《闲情偶寄·器玩部》，清康熙十年（1671）刻本，目录第 1 页

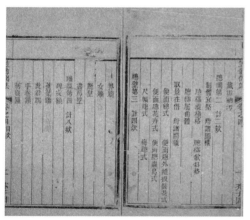

图 2-23　清·李渔《闲情偶寄·器玩部》，清康熙十年（1671）刻本，目录第 2 页

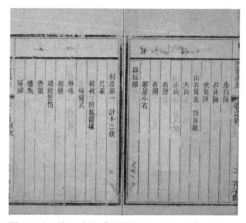

图 2-24　清·李渔《闲情偶寄·器玩部》，清康熙十年（1671）刻本，目录第 3 页

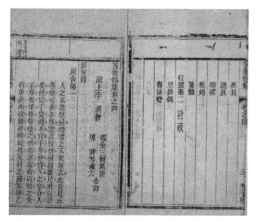

图 2-25　清·李渔《闲情偶寄·器玩部》，清康熙十年（1671）刻本，目录第 4 页

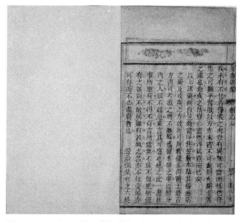

图 2-26　清·李渔《闲情偶寄·器玩部》，清康熙十年（1671）刻本，正文页

李渔通过茶壶和酒壶不同功能的设计，说明置物要"取其适用"，从方便使用者的需要出发，从各种壶的不同储存对象考虑，从怎样使流水畅通的关系来设计壶的造型和结构。基于人们的文化习俗，他提出茶壶设计"凡制茗壶，其嘴务直"的设计原则。这种源于生活、善于观察、因材施艺、注重器物使用功能与实用价值的设计思想值得后人推崇。

Li Yu has demonstrated that we should choose the most appropriate part of a thing by means of different designs of functions to the teapot and flagon. Considering the actual needs of users, different pots and storage materials and relationship of how to make water flow smoothly, the designers ought to design the proper models and structures of pots. On the basis of the people's cultural customs, he has put forward the design principles targeting to the teapot making, for example "the spout must be straight when making teapot". These designing ideas considering both the using functions and practical values which would originate in the daily life, careful observations and make designs in accordance with different materials deserve the learning of later generations and praise highly.

《闲情偶寄》是清代李渔①撰写的一部综合性著作。全书分为词曲、演习、声荣、居室、器玩、饮馔、种植、颐养八部。书中居室部、器玩部集中论述了房舍构筑、窗栏、墙壁、楹联、匾额、山石和床帐、几榻、橱柜等的构式制作与陈列，他主张朴素节俭，反对奢侈夸张；主张精巧适宜，反对徒有其表；主张创造出新，反对仿效雷同。充分体现出作者对居室家园装修的独居匠心之见解。

The comprehensive book titled *Xian Qin Ou Ji* is written by Li Yu in Qing Dynasty. The whole book consists of the eight sections such as a general term for words & music, exercise, reputation (声荣), residence, enjoyable implement (器玩), catering culture (饮馔), plantation, and health maintenance. In the book, the residence and enjoyable implement mainly discuss various structural production and display, such as construction of houses, window lattice and bar, wall, couplet plaque, hill stone, sparver (床帐), couch and cabinet. He advocates simplicity and frugality, and opposes luxury and exaggeration; advocates delicacy and appropriation, and opposes good look without substantial ability; advocates innovation and creativity, and opposes simulation and similarity. All of these fully embody the author's ingenious opinions on home decoration.

---

① 李渔，字笠鸿，号笠翁，浙江兰溪人。他著作颇丰，传承至今主要有小说集《李笠翁曲话》《十二楼》《笠翁十种典》《资治新书》以及诗文《一家言》等。

2.8 清·钱泳《履园丛话·造园》(节选)

Lv Yuan Cong Hua. Author Qian Yong. Qing Dynasty (Selected Sections)

"造园如作诗文，必使曲折有法，前后呼应，最忌堆砌，最忌错杂，方称佳构（图 2-27~图 2-30）。"

园林建造与吟诗撰文如出一辙，必定能使造园曲径通幽、开合收放、曲折含蓄、前后呼应。建造园林最要戒除垒积砖石的直白空间，最忌讳杂乱无章的布局，才能称得上引人入胜的园林建造。造园和写诗两者都需要经过构思，互相交错但条理分明、井然有序、富有韵律；无论是园林的布局，还是诗文的字句都需要前后相接、首尾呼应，使得转承顺理成章；造园如同写诗一般最忌讳使用"华丽无用"的辞藻叠加而成，最忌讳交错混杂而缺乏头绪，唯有攻克这些难处，才能称为上乘的杰作。

此段的关键在于古代园林建造何以称为"佳构"的方法，中国古典园

Landscape architecture paralleles reciting poems and writing articles, which would be bound to make gardening tortuous, implicit, open and close, a winding path leading to a secluded spot, and act in cooperation. Landscape architecture should be mind of avoiding the blank space made by stones and bricks and the layout in a mess. If so, it is qualified to be called an attractive and fascinating garden. The landscape architecture, like the poetry creation, should be coherent, logical, rhythmic, and well-regulated in good order through ideas. No matter the garden layout or the poetry writing both need the carrying and inclusion which could enable the shift to be reasonable. The garden buildings have to avoid being mixing and illogical, which could be the same as the poetry writing that loads one's work with too many fancy phrases. Only by tackling these difficulties can they be called superb masterpieces.

The key to this paragraph is the method of how to judge ancient landscape building as architectural masterpieces. The

图 2-27 清·钱泳《履园丛话·造园》,清道光五年(1825年)述德堂刻本,封面

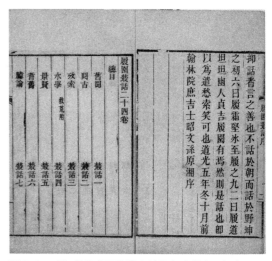

图 2-28 清·钱泳《履园丛话·造园》,清道光五年(1825年)述德堂刻本,目录第 1 页

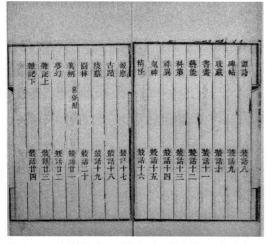

图 2-29 清·钱泳《履园丛话·造园》,清道光五年(1825年)述德堂刻本,目录第 2 页

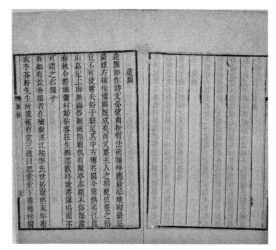

图 2-30 清·钱泳《履园丛话·造园》,清道光五年(1825年)述德堂刻本,正文页

林的造园艺术蕴涵了丰富的中国文化思想,体现古代文人对音乐、文学、书画的品鉴,运用山水花木、亭台楼阁、池馆水榭组合而出,如诗如画、妙不可言。"造园如作诗文",启迪人们对我国古代造园艺术的思考。我国的古典园林反映的是文人雅士的审美

construction arts of classical Chinese ancient gardens contain rich Chinese cultural ideology and reflect the ancient scholar's appreciation of music, literature, painting and calligraphy. These could combine mountains, rivers, flowers, trees, pavilions, terraces, open halls and waterside pavilions, which would feel poetic and picturesque too wonderful for words. To build a garden is like to write poetry and article. It could inspire and

品位、能工巧匠的精湛技艺，自然景观与诗、书、画的意境交相辉映，融合在园林实景之中，中国文化思想与造园设计具有异曲同工之妙。

　　钱泳的《履园丛话》是一部清代运用笔记形式撰写而成的著作，全书共24卷，翔实地记述了作者的亲身经历，内容丰富、门类繁多，各类记载均指明源出。本篇《造园》是第二十卷《园林》中的一篇叙事文，不仅提出我国古代造园的要旨和园林建造优劣的鉴赏标准，并且体现我国古人崇尚造园文化，可谓"造园兴游以成诗文"。明代的计成、文震亨，清代的李渔、袁枚等名家都有所著述，他们能文善诗、书画兼长的文化传统皆体现在造园之境界中，园林在我国古代已成为文人雅集的胜地。可见，中国古典园林艺术独树一帜，它与中国传统文化相辅相成、同源共流。

enlighten our thoughts of Chinese ancient garden arts. The classical Chinese gardens embody the aesthetic taste of refined scholars and the excellent techniques of skillful craftsmen. The natural landscape adds radiance and beauty to the atmosphere of poem, calligraphy and painting, which are all compromised in the scenes of garden. Chinese cultural ideas could achieve the same goal with garden construction in a certain way.

Lv Yuan Cong Hua written by Qian Yong is a type of short sketches monography in Qing Dynasty. The book has twenty-four scrolls and precisely describes the author's personal experience, richness in content, variety in ranges, and records specifies sources. This text titled *Gardening* is a narrative text from the twentieth scroll of *Landscape Architecture*. It not only puts forward the gist of gardening and appreciation standards of ancient Chinese gardens construction but also reflects that the ancient Chinese advocated gardening gem, which can be described as "composing a poem after visiting gardens under construction." These masters such as Ji Cheng, Wen Zhenheng in Ming Dynasty and Li Yu, Yuan Mei in Qing Dynasty composed their writings about gardens, and the atmosphere of gardens fully demonstrated their highly abilities in literate, writings and paintings. Garden became a famous scenic spot in ancient times for refined scholars. It can be seen that the art of Chinese classical gardens are unique, which are complementary with the Chinese traditional culture and harmonious with same sources and originality.

思考题：

　　1. 请说出"斗栱""造园""园屋"的专用语表达。
　　2. 请用双语表述的方式，简述中国古典文献中蕴含的"设计"理念。
　　3. 根据古典文献节选和附录一的内容，列举出三部论及设计文化的典籍。

第 3 章　中国近代设计文化概论
Chapter Three: Topic on Chinese Design Culture in Modern Times

[本章导读]

20世纪初期中国的设计教育家、美术教育家大多曾有海外留学经历，他们以间接或者直接的方式接受西方艺术与设计体系，回国以后，这一批留学生办学施教、撰写图案教材以及西方设计史论书籍，将他们认为先进的、最适合中国国情的西方艺术设计与教育体系传入中国，并且以学贯中西的独特的设计教育思想及方法因材施教，积极推进中国设计与美术领域的发展。

本章列举20世纪初期的八位设计教育家，十部图案与装饰教材，这些史料真实再现了当时中国最前沿的设计状况。此时，中国设计领域曾试以西方文化作为参照物，一方面尝试借鉴西方设计文化，由外向内实施设计革新，一方面尊崇先人传承中华

3.1 先导：中国设计启蒙的教育家
Forerunner: Chinese Enlightened Design Educator

20世纪初期，中国教育领域的革新以新式学堂和留洋归国学者两支主力组成。新式学堂以培养技能型教育人才为开端，全国各地出现图画课、手工课并且学生已开始在学堂接受西方艺术课程与教学模式的熏陶。曾在欧美及日本学习的前辈中国学者们，早已闻知包豪斯及其对西方设计的广泛影响，但是，他们并没有急于将当时盛行于欧洲的包豪斯设计美学思想、观念与技术直接引进中国，而是将西方美学体系的"装饰、图案、图画手工"等核心内容传播至中国。他们逐渐地将西方设计教育体系引入中国新式学堂，涌现出一批我国近代设计教育的启蒙教育家。

The new-style schools and the oversea returned scholars were two main components for the field of Chinese education innovation in the early 20th century. At the beginning, there were drawing curriculum and manual curriculum to cultivate training skilled talents in new-style schools all over the country. Students had begun to accept the influence of Western art curriculum and teaching model. The older Chinese scholars who studied in Europe, America and Japan had already heard the Bauhaus and extensive influence on Western design but they did not rush to directly introduce prevailing European design aesthetic ideas, concepts and techniques of Bauhaus into China. They selected to pass on the core content of Western aesthetic system such as decoration, pattern, and manual drawing into China. They gradually introduced the Western design education system into Chinese new school and emerged a group of enlightenment educator of modern design education in our country.

一、陈之佛
I. Chen Zhifo

陈之佛（1896—1962），出生于

Chen Zhifo (1896 – 1962), the well-known Chinese

浙江余姚，中国著名美术教育家、国画家、工艺美术家。1918 年他曾留学日本，入东京美术学校工艺图案科学习，1923 年回国后创办上海图案馆，担任上海东方艺专图案科主任、《东方杂志》等书刊装帧设计工作，随后担任广州美专图案科主任，首开我国近代图案作品展之先河，为我国设计教育发展奠定基础，相继出版了《图案教材》《图案 ABC》《图案构成法》等融会贯通中西方设计方法的教材与专著。

陈之佛编著的系列教材系统总结了图案的样式、种类、变化规律以及学习的用途与意义，其教学理念自然体现了"中西合璧"的设计思想。他精通中国传统文化并擅长工笔花鸟画，是将宋、元以来中国画传统技法、西画方法与图案规律融会贯通的工笔花鸟画大师，他的作品展现出丰厚的中国传统文化印记；与此同时，他致力于西方图案教学研究，他将凝聚着西方文化因素的图案设计引入中国设计教学领域，并能够结合中国国情传播西方设计文化与教学方法，可谓中西兼长、取其精华。

art educator, traditional Chinese painter and craft artist, was born in Yuyao, Zhejiang Province. In 1918, he studied at Tokyo National University of Fine Arts and Music in Japan, majoring in design and crafts. Returning to China in 1923, he established the Shanghai Art Pattern Design Museum and acted as the director of the Pattern Design Department in Shanghai Oriental Academy of Fine Arts. In addition, he took charge of the graphic design for the *Journal of the Orient* and other books and periodicals, and later served as the director of the Pattern Design Department at Guangzhou Academy of Fine Arts. As the pioneer of Chinese modern art exhibitions, Chen Zhifo published many textbooks and monographs, displaying a thorough understanding of the design methods through mastery of Chinese and Western design, such as *Pattern Design Arts, ABC of Pattern Design Arts*, and *Principles of Pattern Design Structure*, which combined to lay a solid foundation for the development of domestic design education.

The series of teaching material compiled by Chen Zhifo systematically summarized the styles, categories and changing rules of pattern design, and also the usage and significance of studying pattern design art. The ideas behind the series naturally reflected the concept of "a combination of Chinese and Western elements." He had an intimate knowledge of traditional Chinese culture and was a master in meticulous flowers-and-birds paintings. He was also able to merge traditional Chinese painting techniques since the Song and Yuan Dynasties, Western painting techniques and pattern design techniques. His works were marked by profound imprints of traditional Chinese culture. Chen Zhifo devoted himself to studying Western pattern design teaching, and aimed to introduce Western cultural elements of pattern design into the Chinese design education (field). In so

在设计实践方面，陈之佛作为中国早期设计实践者之一，他大胆借鉴西方历史文化元素并运用于设计作品之中，他设计的《东方杂志》期刊封面第二十四卷第十一号（1927年6月10日出刊）、第二十四卷第十二号（1927年6月25日出刊），设计风格简洁明快、构图新颖、视觉冲击力较强，从构成形式到视觉符号显然受到20世纪20年代西方装饰艺术运动及其装饰风格（Art & Deco, 1920-1930）的影响（图3-1、图3-2）。

二、雷圭元

雷圭元（1906—1988），上海松江人，出生于北京，著名工艺美术设计家及教育家。21岁毕业于国立北平艺术专科学校并留校任教，23岁曾赴巴黎留学，25岁回国，先后任教于国立杭州艺专、中央美术学院、中央工艺美术学院等。著有《工艺美术技法讲话》《新图案学》等大学适用教材。

雷圭元基于图案理论研究，注重对中国传统图案的重新整理与定

doing, he disseminated Western design culture and teaching methods in accordance with our national characteristics and managed to absorb and give full play to the quintessence of both cultures.

As one of the earliest domestic design practitioners, Chen Zhifo in the practice of art design boldly took references from Western historical and cultural elements and applied them into his art and design works. He designed covers of the Journal of the Orient for the Vol. 24, No. 11 published on June 10th, 1927 and Vol. 24, No. 12 published on June 25th 1927. His design style was of great concision and clarity, with novel composition and strong visual impact. Their composition form and visual symbols were obviously affected by the Western decorative art campaigns and decorative styles in 1920s (Art & Deco, 1920-1930) (Illustration 3-1, Illustration 3-2).

II. Lei Guiyuan

Lei Guiyuan（1906-1988), the distinguished craft designer and educator was born in Beijing and originally from Songjiang in Shanghai. At the age of 21, he taught in National Peking Art School after graduation. He went to study in Paris when he was 23 years old and returned to China at the age of 25. Since then, he had been teaching in National Art College of Hangzhou, China Central Academy of Fine Arts, and Central Academy of Art and Design, etc. He wrote some textbooks including *New Pattern Arts and Talks on Arts and Crafts Techniques* for university students.

Based on pattern design theory research, Lei Guiyuan laid emphasis on the rearrangement and redefinition of

第3章 中国近代设计文化概论
Chapter Three: Topic on Chinese Design Culture in Modern Times

图 3-1 陈之佛设计的《东方杂志》封面，第二十四卷，第十一号，1927 年 6 月 10 日出刊

图 3-2 陈之佛设计的《东方杂志》封面，第二十四卷，第十二号，1927 年 6 月 25 日出刊

义，提炼与概括中国传统图案中的几何形象与构图规律，总结出图案设计中平视体、立视体、格律体、"回"形图案等形式法则，他的图案教学模式中西融合、由简入繁、循序渐进，阐释了独到的见解，有助于启发学生的设计思维，具有独创精神。当时，他已意识到"图案"教学作为实用美术的重要作用，希冀推广实用美术（即设计教育）以及革新中国艺术的想法是根深蒂固的，艺术应反映时代精神，博采众长，体现

traditional Chinese pattern design. He tried to refine the geometrical figures and composition rules of traditional Chinese pattern art, and summarize form principles such as isometric composition, perspective composition, metrical composition and pattern art in the shape of character "Jiong". His pattern design teaching mode was a fusion of Chinese and Western characteristics, growing from scratch to specific and proceeding in an orderly way. His original views, which could contribute a lot to inspiring students' design thought, showed his creative spirit. At that time, he already realized the important role of "pattern" teaching as a practical arts, his idea of promoting practical art, the design education and

时代进步之急需，不但要有专业精深的美术教育，更需要顺应社会发展、美化环境、提升人们生活水平的实用型设计教育。

值得一提的是，雷圭元也为中国现代漆画带来一片新气象，他突破旧文化观念的束缚，将欧洲新型漆画艺术表现方法与中国传统漆画工艺结合，将纯美术的审美观念融入平面的漆绘艺术，形成中西多元化方法的融合、多种材质的并用，使传统的中国工艺走向创新之途。

reforming Chinese art was ingrained. The art ought to reflect the spirit of the times, learn widely from others' strength, present the urgent need of time progress. Our Society not only wanted professional and profound art education, but also wanted practical design education, which could follow social development, improve beautiful environment and improves people's living standards.

And it is worth mentioning that Lei Guiyuan also brought new ideas to Chinese modern lacquer painting. By breaking the shackles of outdated cultural concepts, he combined the expression techniques used in European new-style lacquer painting art with traditional Chinese lacquer painting techniques, and integrated aesthetic idea of fine arts with graphic lacquer painting art, thus leading to the fusion of Chinese and Western techniques, the blending of multiple materials, which would make the innovative progress of traditional Chinese crafts.

三、俞剑华

III. Yu Jianhua

俞剑华（1895—1979），出生于山东济南，我国著名的教育家、史论家、国画家。他20岁考入北京高等师范手工图画专修科，师从陈师曾、李毅士等名师，23岁毕业，次年曾在日本举办个人画展，先后任教于北京美术学校、北京师范学校、山东美术学校、北京交通大学、上海新华艺术专科学校等，主要教授国画和用器画。他撰写出版了《国画研究》《最新图案法》《最新立体图案法》《中国画论类编》《中国山水画的南北宗论》《石涛画语录》等中西艺术论著。

Yu Jianhua (1895-1979), the famous art educator, historian and traditional chinese artist was born in Jinan in Shandong Province. He was admitted to the Beijing Teachers College (the current Beijing Normal University) to study in the specialized subject of manual drawing and art crafts at the age of 20. He studied under the guidance of distinguished teachers such as Chen Shiceng, Li Yishi and so on, and graduated at the age of 23. In the next year, he held his individual painting exhibition in Japan, and then he respectively taught in schools like Beijing Art School, Beijing Normal University, Shandong Art School, Beijing Jiaotong University, Shanghai Xinhua Art School and so on, mainly teaching traditional Chinese painting and instrumental

俞剑华编撰的图案教程，从设计制图的角度来阐释器物制作的法则、器物的结构以及图案装饰的审美观，体现了我国早期设计构思的雏形；从文化根源来看，他受到中国传统绘画艺术的熏陶与涵养，又在北京高等师范手工图画专修科接受了中西结合的艺术教育，当时的学校美术教育已受到西方艺术与设计思潮的影响，他间接地受到西方文化的影响，并始终坚持理论研究与艺术实践相结合的观点，他的设计思想充分体现在他的图案教学法以及绘画创作之中。他不单重视美术专业基础教学，也极为重视人格培养、思想道德以及艺术修养的熏陶。

四、鲁迅

鲁迅（1881—1936），浙江绍兴人，他不仅是我国著名的思想家、中国现代文学的奠基人，又是中国翻译文学的开拓者、中国新兴版画的倡导者。

drawing. He wrote and published works of Chinese and Western arts, such as *Research in Chinese Painting*, *New Pattern Design Methods*, *The Latest Three-Dimensional Pattern Arts*, *Series of Chinese Painting Theories*, *Southern and Northern Theories on Chinese Landscape Painting*, and *Painting Quotations from Shi Tao*.

The Pattern Design Textbooks compiled by Yu Jianhua illustrated the laws of artifacts producing from the perspective of design drawing, structure of artifacts and aesthetic standards of pattern decoration, which would embody the prototype of the early design conception of China. From the view of cultural origin, he was deeply influenced by traditional Chinese painting arts and received the combination of Chinese and Western art education in the specialized subject of manual drawing and art crafts in Beijing Teachers College (the current Beijing Normal University). The art education in school at that time was influenced by Western art and design trends, and he had been indirectly affected by Western culture. He always adhered to the idea of combining theoretical researches and artistic practices, and his thoughts on design had been fully embodied in his pattern design teaching and art works. He not only put emphasis on professional basic art teaching, but also attached great importance to the influence of personality cultivation as well as ideological and artistic moral cultivation.

IV. Lu Xun

Lu Xun (1881-1936), was born in Shaoxing Zhejiang Province. He is not only a distinguished ideologist, but also a founder of the Chinese contemporary literature, a pioneer of Chinese translated literature and an advocator

"他在美学方面专精的研究、对中西美术的传播与普及以及对我国美术教育事业的传承与创新都作出了重要贡献,对当今的版画创作、美术理论和美术教育发展也具有深远影响。"[①] 然而,鲁迅的设计实践对20世纪初期中国设计思想形成具有举足轻重的地位与作用。

标志设计: 1917年鲁迅设计的北京大学校徽

从标志的视觉设计来看,鲁迅的北京大学校徽设计灵感来自中国传统书体之一的"篆书",传承了我国古代自上而下的传统阅读习惯并将"北大"二字作上下叠排,笔画舒展、简练大气、极为巧妙;基于中国传统篆书的构字形态,又将"北"字与"大"字的形态进行拟人化设计,两个汉字呈现上下呼应之势,使其构成元素统一协调;"北"字的设计呈现出对称之态,犹如两个"背靠背"的人像,"大"字设计为一个屹立的正面人像,整体标识具有深刻的意义,展现"三人团结"的设计意象,校徽设计还运用了中国传统印章的圆体造型,结合古代篆书与印章双重的传统文化符号,从而凸显了中国传统文化内涵的人文精神(图3-3)。

of Chinese newly-developing print making. "He made significant contribution to the specialized research on the art aesthetic aspects, the transmission and popularization of the Western fine arts, as well as the inheritance and innovation of the Chinese art education cause, and exerted profound influence on the current print making, artistic theories and development of the art education." However, design practices of Lu Xun had exerted a decisive role and function in the ideological formation of the Chinese design thoughts in the beginning of the 20th century.

Logo Design: Logo of Peking University was designed by Lu Xun in 1917

From the perspective of logo visual design, Lu Xun was inspired by the traditional Chinese calligraphy-seal character in the design of the school badge for Peking University. Furthermore, he had inherited the traditional habit of reading from top to button, and arranged the two characters "BEI" and "DA" in a sequence from up to down. The stretched, concise and elegant strokes were extremely ingenious. Based on the character forming in the traditional Chinese seal character, he implemented unification of the constituent elements and rhythm of the two characters "BEI" and "DA". The character "BEI" manifested the state of symmetry, which would be just like two "back-to-back" figures. The character "DA" was designed as a figure standing tall and upright. They could echo with each other and demonstrate the design image of "three makes a crowd" as a whole, which would embody profound significance and integrates the concerted endeavors

---

① 应宜文,《论文学家鲁迅的〈拟播布美术意见书〉》,载《中文学术前沿》第四辑,2012年5月,第78页。

of Peking University students in enlightening the nation's gems of wisdom with their intelligence. In the design of the school badge, the round outline of the traditional Chinese seal was utilized, which would be evenly proportioned, concise and magnificent. It also combined the dual traditional culture symbols of seal character and calligraphy seal, which could highlight the connotation of the traditional Chinese culture and the profound humanistic quality (Illustration 3-3) .

图 3-3　1917 年鲁迅设计的北京大学校徽

五、陶元庆

陶元庆（1893—1929）浙江绍兴人，我国近代杰出的书籍装帧设计师、画家、美术教育家。他不仅擅长中国画、西洋画和水彩画，更精通于图案设计及书籍封面设计。他最具有代表性的作品是为鲁迅的小说、译著及编著而创作的书籍装帧设计，诸如：《苦闷的象征》(鲁迅著)、《工人绥惠略夫》(鲁迅译)、《中国小说史略》、(鲁迅著)、《唐宋传奇集》、《彷徨》、《朝花夕拾》《一个青年的梦》(鲁迅译)；他最富有特色的作品是为许钦文的小说集而创作的书籍装帧设计，诸如《故乡》、《毛线袜》、《回家》、《幻象的残象》等（图 3-4~图 3-7）。

陶元庆的设计创作犹如中国草书创作一般，往往是在短时间内完成，

V. Tao Yuanqing

Tao Yuanqing (1893-1929), was born in Shaoxing in Zhejiang Province. He is an outstanding book designer, artist and art educator in modern China. He is not only good at Chinese painting, Western painting and watercolor painting, but also proficient in pattern design and book design. His most representative work is book design for Lu Xun's novels, translations and edited books, such as *The Symbol of Dullness* (Lu Xun Author), *Worker Suihui Luefu* (translated by Lu Xun), *A Brief History of Chinese Novels* (Lu Xun Author), *Legend Collection of Tang and Song Dynasty*, *Hesitate at the Cross Road*, *Dawn Blossoms Plucked at Dusk*, and *Dream of A Teenager* (translated by Lu Xun). His most distinctive design work is the book design for Xu Qinwen's novel series, such as *Hometown*, *Wool Socks*, *Back Home*, and *Fancy and Residual Scene* and so on (Illustration 3-4 to Illustration 3-7).

Tao Yuanqing's design forms a coherent whole and often in a short period to complete vividly like traditional

图 3-4 陶元庆书籍封面设计《幻象的残象》之一

图 3-5 陶元庆书籍封面设计《幻象的残象》之二

图 3-6 陶元庆书籍封面设计《工人绥惠略夫》

图 3-7 陶元庆书籍封面设计《回家》

可谓一气呵成，气韵生动。他的书籍装帧设计基本传承了古籍线装的题鉴形式，映射出丰厚的中国传统文化底蕴，一方面，他设计的书籍装帧以图

Chinese general cursive writing. His book design art basically inherited ancient thread-binding inscription and reflected rich Chinese traditional culture accumulation. On the one hand, his artistic book design was dominated by patterns and

案为主的设计形式,注重装饰感与图形语言,将点、线、面结合运用在图像之中,又从传统绘画中受到影响,在设计中大量运用花卉、人物以及传统图案元素,他的设计灵感源自汉代画像砖、六朝画像石,运用灵动的人物图案造型,充满想象力的动物图案造型,又巧用云纹将设计图面贯通一体。另一方面,他的书籍装帧设计形式多样、变化显著,大胆地借鉴西方艺术形式,运用各种西方绘画表现手法,融合了中国画、西洋画和水彩画元素以及纹样的写实变化,并且赋予图案以崭新的形式美感。从他留给后人的设计作品可知,他是我国近代一位中西融合、博采众长又具有独创精神的设计师。这一时期的书籍装帧和封面设计,体现了作者与读者之间的互动与体验,设计风格上杂糅了多元的西方艺术语汇,并沿用中国传统绘画元素,呈现中西交融且富有创意的设计文化风尚。

六、钱君匋

钱君匋(1907—1998),浙江桐乡人,著名书法家、篆刻家、画家、书籍装帧设计家。他的篆刻作品入木三分、收放自如、匠心独运,边款长跋,诗文自撰皆工。他著有《钱君匋印存》一书,万叶楼景印,民国33年(1944年)初版。此书收集了钱君匋的大量篆刻

highlighted artistic decoration and graphic language. His book design integrated points, lines and planes in graphics and was influenced by traditional painting. In his design, flowers, figures and traditional patterns ware frequently used. His art and design inspiration originated from portrait bricks in Han Dynasty and portrait stones in Six Dynasties. His design adopted vivid figures and imaginary animal patterns while using cloud patterns skillfully to integrate design patterns as well. On the other hand, his book designs were various and change dramatically. The designs courageously refered to Western artistic forms and adopted various Western painting techniques. These designs were integrated Chinese paintings, Western paintings, watercolors and realistic patterns and give patterns new aesthetic sense. From his design works for posterity, he was a designer who learned widely from Western and Chinese strong aspects, integrated both styles and keeps creative spirits in modern China. During this period, book and cover design expressed interaction and experience between the author and the reader. His design style absorbed Western diversity arts, evolved traditional Chinese painting elements and presented a blend of Chinese and Western cultural and creative design.

VI. Qian Juntao

Qian Juntao (1907-1998), the well-known calligrapher, seal carver, artist and book designer was born in Tong Xiang in Zhejiang Province. His carving works including seal inscription, colophon, poetic pose and essays entered three-tenths of an inch into the timber – vigorous effort, wield skillfully and had great originality. Written by him, the *Collection of Qian Juntao Seals* was

精品，有别于传统的篆刻字体，风格独特、新颖大胆，字形颇具设计意象（图 3-8~ 图 3-11）。

firstly published by Wan Ye Lou Publishing House in 1944. These seal works collected in his book were different from traditional seal carving, expressed characteristic style, novel and original script with design images（Illustration 3-8 to Illustration 3-11）.

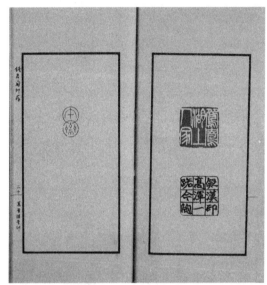

图 3-8 《钱君匋印存》封面（左）

图 3-9 《钱君匋印存》书籍第 20 页篆刻作品（右）

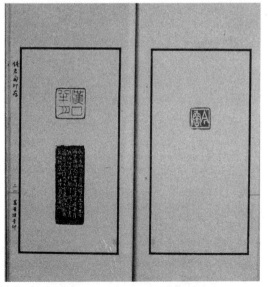

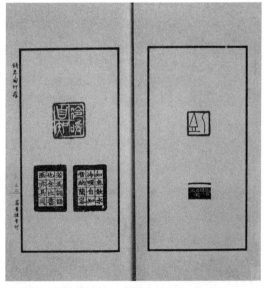

图 3-10 《钱君匋印存》书籍第 21 页篆刻作品

图 3-11 《钱君匋印存》书籍第 32 页篆刻作品

同时，他又著有一部《图案字文集》，新时代书局出版，1932年初版。此书收录了诸多别出心裁而应用广泛的字体设计，富有设计象征功能与意形认知功能。在书结尾部分，刊登了《图案文字集付印题记》，作者提到："这里所收的都是合乎美的法则的，是可以能为图案文字的模范的。只要凭着各人的智慧，有了这一册书，便可无穷尽地从这些已成的字形中寻出新的形来，这原是无限的。"[①]诸如："进行曲选"的字体设计（图3-12）点、线、面排布舒张有序、富有轻快的节奏感，汉字变化成为含义相仿的字形，从而深化了文字的表意功能。又如："高尚趣味"的字体设计特别处理了"点"的形态（图3-13），使之成为菱形的点，增添了字体设计整体的视觉对称效果，又将"趣味"两字进行图案化处理，使之更生动形象，产生欢快的联想。此书以富有变化与表意的字体设计，体现作者对形式美的创新追求与设计构思。

不仅如此，钱君匋还著有两部西方美术史，一部是他的《西洋古

At the same time, his monograph titled *Collected Works of Pattern Style Words* firstly was published by the New Age Bookstore in 1932. This book contained a lot of ingenious and widely used character font's design, which would have rich design symbolic function and ideology cognitive function. In the end of the book, the "Explanation Notes on the Collected Works of Pattern Style Words before pressing" published, and the author mentioned, "Here included are in line with rules of beauty, which could be a model of pattern character. As long as the wisdom of individuals with this book, they could find new shapes from these already formed fonts endlessly. It has been infinite."[①] Taken the "Selection of March" font design as an example (Illustration 3-12), the point, lines, and plane could have been arranged relaxed and orderly, and full of brisk rhythm sensation. Chinese characters were changed into a character pattern with similar meaning, thus deepening the ideographic function of characters. As another example, in the "Lofty Preferences" font design, it could deal with the form of "points" especially (Illustration 3-13), which would make it as a rhombic point, and add the overall visual symmetry effect of fonts design. It also handled patterning process to two words of "fun", which would make it more vivid and produce a happy association. This book contained a lot of ingenious and widely used character font's design, which could have rich design symbolic function and ideology cognitive function. This book expressed the author's innovation pursuits and design concepts through varied and ideographic character font design.

Not only that, Qian Juntao also published two monographs of Western Art History. One was his *Western*

---

① 钱君匋编，《图案字文集》，新时代书局出版，1932年出版，第140页。（Qian Juntao (ed.) (1932) *Collected Works of Pattern Style Words*. New Age Bookstore. p.140）

图 3-12 钱君匋编著《图案字文集》中"进行曲选"字体设计

图 3-13 钱君匋编著《图案字文集》中"高尚趣味"字体设计

代美术史》,1946 年 6 月初版,永祥印书馆(全书 79 页),论及从西方旧石器时代、新石器时代之美术、埃及的古代美术至 19 世纪前期的古典主义、浪漫主义美术史。另一部是他的《西洋近代美术史》,1946 年 9 月初版,永祥印书馆(全书 72 页),阐释从西方的写实派、自然派、新理想派、印象派至现代法国的工艺美术、北欧现代的工艺美术。可见,钱君匋对西方美术史以及由西方传入的应用字体设计颇有研究,他的篆刻作品基于中国传统文化的深厚功底,融会了西方字体设计体系之美的法则,实为兼收并蓄、融合创新之佳作。他不愧为一位一专多能的艺术家(图 3-14、图 3-15)。

Ancient Art History, firstly published by the Yong Xiang Printing House in June 1946. The book did have 79 pages in total, mainly expounded from the Western Paleolithic, Art of the Neolithic Age, Ancient Egyptian Art to the classical, romantic art history in early nineteenth century. The other one was his *Western Modern Art History*, firstly published by the Yong Xiang Printing House in September 1946, 72 pages in total, which would expound from the Western Realism, Naturalism, Neo-idealism, and Impressionism to the modern French arts and crafts, the Nordic modern arts and crafts. It can be seen that Qian Juntao had a lot of research on Western art history and application front design transmitted from West. His seal carving works based on the deep foundation of Chinese traditional culture, combing with the rules of the beauty of Western typography, were true "inclusive" and innovative masterpieces. He did deserve to be a highly-qualified and multi-dimensional artist (Illustration 3-14, Illustration 3-15).

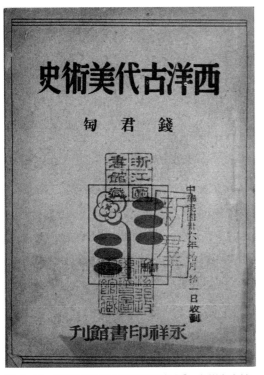

图 3-14 钱君匋著《西洋古代美术史》,永祥印书馆,民国 35 年(1946 年)六月初版

图 3-15 钱君匋著《西洋近代美术史》,永祥印书馆,民国 35 年(1946 年)九月初版

七、傅抱石

傅抱石(1904—1965),江西新余人,在中国画领域成就卓著,他是一位公认的国画家,著有《山水人物技法》《明末民族艺人传》《傅抱石画集》等,也是一位艺术理论家,著有《木刻的技法》《中国绘画理论》《中国美术年表》《中国绘画变迁史纲》等。

VII. Fu Baoshi

Fu Baoshi(1904-1965), was born in Xinyu in Jiangxi Province. He is an acknowledged traditional Chinese painter. He made brilliant achievements in the field of traditional Chinese painting. His representative works could include all such as *Techniques of Landscape and Figure Painting*, *The Biography of National Artists in Late Ming Dynasty and Art Creation of Fu Baoshi*. At the same time, he was also a famous artistic theorist and composed many brilliant academic works such as *Techniques of Wood Engraving*, *Theories of Chinese Painting*, *Chronology of Chinese Art*, and *The Development History of Chinese Painting*.

然而，傅抱石也曾热衷于西方图案设计的编译、教学与研究，撰写了《基本图案学》和《基本工艺图案法》两部反映设计法则与形式美要素的教材。这是两部体系完备的图案设计的教科书，正如《基本工艺图案法》一书中提出："图案之原则即美之原则也。"可见，图案凝练了造型、式样、色彩、装饰等各种美的要素。他的教学理念自然体现了"中西合璧"的设计思想。

傅抱石不仅擅长国画山水和人物，将唐、宋以来传统中国画用笔、用墨技法与图案艺术规律融会贯通，还精通中国传统文化，其作品展现了深厚的中国传统文化底蕴。他提倡的美术教育目标不局限于传授与掌握中西绘画技法，而是积极推行实用美术教育，并提出"发扬中华文化"较为宏观而高远的美术教育宗旨，他的教学理念富有前瞻性，也是我们每一位热爱中华文化的人所期盼的。与此同时，他致力于西方图案教学研究，中西兼长、取其精华，他通过编译图案教材等途径，结合中国国情传播20世纪初期的西方设计文化与教学方法，为创建中国早期设计教育体系作出很多贡献。

In addition, out of strong fascination with the compilation, teaching and research of Western arts, Fu Baoshi also composed two textbooks that could reflect design guidelines, methods and formal aesthetic elements, namely *Basic Pattern Studies and Guidelines of Basic Crafts and Pattern Design*. These were two systematic and contextual textbooks for vocational schools. As mentioned in the book of *Guidelines of Basic Crafts and Pattern Design*, "the principle of pattern design is the principle of aesthetics." This obviously showed that pattern design actually condensed the elements of aesthetics including modeling, style, color and decoration. His teaching ideas naturally reflected the concept of "a combination of Chinese and Western elements".

Fu Baoshi was good at Chinese Landscape and figure paintings and well versed in traditional Chinese painting, techniques of ink and artistic pattern regularity since Tang and Song Dynasties, but also a master of traditional Chinese culture. His art work showed deep Chinese traditional culture inside. His advocated art education target was not only to pass on teaching and mastering the skills of Chinese and Western painting, but also promote practical art education actively. He also proposed relatively macroscopic and lofty art education principle of "carrying forward Chinese culture" and his philosophy of teaching concepts was with prospective, which could also be expected by people who had a keen sense of Chinese culture. At the same time, he devoted himself into the study of Western pattern teaching, was expected in Chinese and Western arts, achieved a mutual development and absorbed the quintessence. He introduced Western design culture and teaching methods combined with China's actual conditions in the early 20th century through translating and editing pattern teaching textbooks and other ways. He made a lot of contributions for establishing the early design education system in China.

八、颜文梁

颜文梁（1893—1988），出生于江苏苏州，我国著名的美术教育家、油画家。自幼随父学习中国传统花鸟画，16岁时考取上海商务印书馆学习刻印、制版和印刷技术，随后进入图画室学习西洋画，这是他最早接触西方艺术与西洋画的经历。1928年他赴法国留学，考入法国巴黎高等美术专科学校学习油画，其间曾考察西欧各国，1932年回国，主持苏州美术专科学校的教学，主要成就在于参照西方美术院校教学方法改造苏州美专，使该校成为20世纪30年代著名的私立美术学校，他提出"中西合璧，造就人才"的办学方针。

颜文梁不愧为我国设计教育与实践的先导者、探索者。1922年，他创建苏州美术专科学校并于1928年赴法国留学。这个时期，他间接地向西方学习，提出苏州美专的教学方针"中西绘画兼学"，并提出了'忍、仁、诚'三字作为苏州美专的'校训'"。① 这为他日后推行设计教育奠定了基础。1932年，他学成回国致力于教育改革与实践。

VIII. Yan Wenliang

Yan Wenliang (1893-1988), was born in Suzhou, Jiangsu Province. He was a well-known Chinese art educator and oil painter. He had studied traditional Chinese flower and bird paintings from his father since childhood. He was admitted to the Commercial Press of Shanghai to study technologies of carving, platemaking and printing at the age of 16. Later on, he entered the drawing studio to study Western painting. It was the first experiences that he had access to Western art and Western painting. He was admitted to Paris Superior Fine Arts School of France to study oil painting in 1928. During the period, he visited Western European countries. He returned to China in 1932 and presided over teaching of Suzhou Academy of Fine Arts. His major achievements included: improving Suzhou Academy of Fine Arts according to teaching methodologies of Western art institutes, making the academy become a well-known private art school during 1930's, and proposing the school running guidelines as follows, "combination of Chinese and Western teaching methods to cultivate talents".

Yan Wenliang deserves such titles as pioneer and explorer of China's design education and practice. In 1922, he founded the Suzhou Academy of Fine Arts and went to study in France in 1928. During that period, he indirectly learned from the west and proposed the teaching guidelines of "Study of Both Chinese and Western Paintings" for Suzhou Academy of Fine Arts. Besides, he put forward the motto of Suzhou Academy of Fine Arts such as "tolerance, benevolence and integrity". ① It laid a foundation for his promoting design

---

① 陈瑞林. 20世纪中国美术教育历史研究. 北京：清华大学出版社，2006：99.（Chen Ruilin.（2006）*Fine Art Education in the 20th Century China: A Historical Perspective.* Beijing: Tsinghua University Press，2006. p99.）

为了解决中国当时许多美术作品、画册和商业图片不得不送往国外制版和印刷的问题,"1933年,学校开办实用美术教学,并设印刷制版工场,由朱士杰教授主持。颜文梁曾亲赴上海订购各种仪器及印刷机器,聘请印刷技师。"①印刷技术的革新,为苏州美术专科学校的应用型美术教学发展提供了必要条件。这是第一所设置"实用美术科及印刷制版工场"的中国美术学校,学校的美术教育革新体现在将美术教学与实际生产相结合,大力推行实用美术教学课程体系,培养社会急需的实用型美术人才,我国第一批印刷制版人才就出自这所学校。

颜文梁还在校内积极建立实用美术科,相当于设计科,以其独特的方式推行"生产工场"的美术教育改革是国内首创的,这与21世纪现代美术院校实施的"艺术工作室制"具有殊途同归的教学目标。可见,苏州美专实施的教学改革实践成为我国实用美术以及设计教育的开蒙,颜文梁也是我国设计教育启蒙的先导之一。

education later on. In 1932, he returned to China and was dedicated to education reform and practice.

At that time, plenty of artistic works, albums and commercial pictures had to be sent abroad for platemaking and printing. In order to solve this problem, he conducted reform in printing technology which provided necessary conditions for development of applied art teaching in Suzhou Academy of Fine Arts. "In 1933, the school set up functional art education and opened a workshop of printing and platemaking which would be presided by Professor Zhu Shijie. One time, Yan Wenliang had been to Shanghai in person for not only ordering varied instruments and printing machines, but also inviting printing technician."① It was the first Chinese art school which set up an "Applied Art Department and Printing Platemaking Workshop". The school's art education reform was demonstrated in combination of art teaching and practical production, and made great efforts to implement the curriculum system of practical teaching of arts and cultivate applied art talents urgently needed in the society. China's first batch of printing platemaking talents graduated from the school.

Besides, Yan Wenliang proactively established an applied art department on campus which was technically a design department. He promoted the art education reform of "production workshop" in a unique way, which would be pioneering in China. It set up the same teaching objectives with "art studio" implemented by fine art academy and institutes in the 21th century. It can be noted that the teaching reform carried out by Suzhou Academy of Fine Arts enlightened China's applied art and art education. At the same time, Yan Wenliang became one of China's design education enlightenment forerunners.

---

① 陈瑞林. 20世纪中国美术教育历史研究. 北京:清华大学出版社, 2006:101.(Chen Ruilin.(2006) *Fine Art Education in the 20th Century China: A Historical Perspective*. Beijing: Tsinghua University Press, 2006. p101.)

3.2 先锋：中西合璧的设计教材
Pioneer: Design Textbooks Combining of Chinese and Western Styles

20世纪初期，随着西方艺术思潮及其设计风格陆续传播到中国，诸如：新艺术运动（Art Nouveau, 1890~1914）、装饰艺术运动及其装饰风格（Art & Deco, 1920~1930）、立体主义艺术（Cubism, 1907~1914）、未来主义运动（Futurism, 1909~1944）、超现实主义运动与达达主义（Surrealism & Dadaism, 1919~1939）等，20世纪30年代，中国悄然兴起一股学习西方的图案设计、版式设计、装帧设计、广告设计、建筑设计等的热潮。为了迎合人们对于西方设计的关注与兴趣，知名的出版社相继出版了各种的图案教材，诸如:《图案教材：中等学校适用》(（图3-16、图3-17)《普通平面图案画法》、《实用图案画法》、《图案画法》、《现代工艺图案构成法》（1~3编）等。这些书籍的作者基本都是从海外留学回国的设计教育家和艺

At the beginning of 20th century, an upsurge of learning Western pattern design, format design, graphic design, advertising design, and architectural design emerged quietly in China in 1930s. Western art concepts and design styles spread to China successively, for example the Art Nouveau, Art & Deco, Cubism, Futurism, Surrealism & Dadaism, and so on. In order to cater to people's attention and interest on Western design, famous publishing presses published various teaching textbooks of pattern design in succession, such as the *Pattern Textbook: Applicable to Secondary School* (Illustration 3-16, Illustration 3-17), *Techniques of General Graphic Pattern*, *Applied Pattern Techniques*, *Technique of Pattern Design*, *Composition Methods of Modern Technological Patterns*（1-3）. The authors of such books were design educators and artists who would basically have the experience of studying abroad and come back abroad. This book did research from ten design textbooks which would have wide influences in the early 20th century as following.

术家。本教程共梳理与研究了十部20世纪初期影响力较广的设计教材，具体如下。

第一部：《图案教材》

《图案教材》（图 3-16、图 3-17），陈之佛编著，上海天马书店1935年初版。全书293页，共18部分，即图案的目的与意义、种类、图案资料与模样化、图案的均齐与平衡、单独模样、轮廓模样、二方连续模样、四方连续模样——散点法、四方连续模样——连缀法、四方连续模样——重叠法、四方连续模样——以一单位排列变化而构成的方法、几何形的模样、地与纹同形的模样、风景的图案、表号的图案、文字的图案化、小品装饰图案、工艺品图案，附有图例780图。书前凡例：本书的编制以适合中学教材为目的，师范学校职业学校均可适

The First Textbook：*Lectures of Pattern Design*

Written by Chen Zhifo, (Illustration 3-16, Illustration 3-17) was published by TIANMA Bookstore in 1935. It has 293 pages and is made up of eighteen sections, including purpose and significance of pattern design, categories, pattern design references and patterning, pattern design uniformity and balance, single pattern, outline pattern, consecutive double-square pattern, consecutive four-square pattern-scattering method, consecutive four-square pattern-clustering method, consecutive four-square pattern-overlapping method, consecutive four-square pattern-single unit permutation and variation, geometric pattern, same-shape of ground and grain pattern, pattern design of landscape, pattern design of numbers, pattern design of words, simple decorative pattern, and pattern of craft arts, which would attach up to 780 pictures. The preface showed a note 'This book is compiled for the purpose of fitting the teaching

图 3-16　陈之佛编著《图案教材》，上海：天马书店，1935年，封面

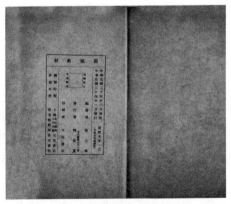

图 3-17　陈之佛编著《图案教材》，上海：天马书店，1935年，版权页

用,务求新颖、切于实用,并能启迪学生之兴趣。

第二部:《工艺美术技法讲话》

《工艺美术技法讲话》(图3-18),原著雷圭元,南京正中书局1936年初版。全书201页,书前为著者自序及作品《浴后》(蜡染),共八章,分别是蜡染技法讲话、夹板及缝纹染技法讲话、三种型纸印染法、天然漆与人工漆的装饰技法讲话、嵌玻璃窗饰装饰技法。每章节前附上著者作品范图,分别展示不同工艺制作法的作品特点与效果。封面标注:国立杭州艺术专科学校艺术丛书。

materials in middle schools, applicable for both normal schools and vocational schools, striving for novelty, originality, and practicability, as well as enlightening students' interests.'

The Second Textbook: *Talks on Arts and Crafts Techniques*

The first edition of *Talks on Arts and Crafts Techniques* (Illustration 3-18) was written by Lei Guiyuan author and published by Nanjing Zheng Zhong Press in 1936. The book has 201pages and eight chapters, and starts with the author's preface and his work *After Shower* (batik). The book involves talks on Batik Technique, Dyeing Technique Combining Splinting and Suturing, Three Kinds of Paper Stencil Printing and Dyeing, Decoration Techniques of Natural and Artificial Paint, and the Decorative Techniques of Window Decorations Embedded in Glass. Author's works are attached at the beginning of every chapter, demonstrating the features and effects of different craft techniques. The cover of the book is marked with "Series of Art Books of National Art College of Hangzhou".

图3-18 雷圭元著《工艺美术技法讲话》,南京:正中书局,1936年

第三部：《最新立体图案法》

《最新立体图案法》（图 3-19）编者俞剑华，上海商务印书馆 1929 年初版。书前编者自序，共五章，分别为总论、器物形状组成法、尺度律、制作器物之方式、装饰法、器体各部概论，书内附有范图解说。

第四部：《图案教材：中等学校适用》

《图案教材：中等学校适用》（图 3-20、图 3-21）郑棣、方炳潮撰，浙江正楷印书局 1936 年出版。全书 76 页，以课程编目，分别为概论、植物变化、动物变化、人体变化、文字变化、适合模样、二方连续、四方连续、角花图案、枕头图案、风景图案、封面图案、标题装饰图案、广告图案、礼品图案、地毯图案、伞子图案、灯罩图案、瓷器图案共十九课。书前郑棣、方炳潮作序二篇，首页例言指明"遵照新课程图案课程标准编辑"，作为中等学校图案教材及无师自修用书。此书特点切于实用、精练概要。

The Third Textbook: *The Latest Three-Dimensional Pattern Arts*

The book titled *The Latest Three-Dimensional Pattern Arts* (Illustration 3-19) was edited by Yu Jianhua and published by the Commercial Press of Shanghai in 1929. This book has five chapters including Author'S Preface, General Introduction, Compositional Approaches of Aifact Shapes, Scaling Laws, Approaches of Making Artifacts, Decoration Approaches, Introduction to All Parts of Artifacts and Picture Explanations.

The Forth Textbook: *Pattern Textbook: Applicable to Secondary School*

Pattern Textbook: Applicable to Secondary School (Illustration 3-20, Illustration 3-21): written by Zheng Di, Fang Binchao, published by the Zheng Kai Press of Zhejiang in 1936, catalogued in courses, composed of nineteen courses including Introduction, Changes of Plant Pattern, Changes of Animal Pattern, Changes of Human Body Pattern, Changes of Script Pattern, Suitable Decorative Pattern, Two-Dimension Series, Four-Dimension Series, Pattern Design for Corner Decoration, Pattern Design for Pillow Decoration, Scenery Pattern, Pattern Design for Cover Decoration, Patterns of Title Decoration, Pattern Design for Advertising, Pattern Design for Gifts, Pattern Design for Carpets, Pattern Design for Umbrellas, Pattern Design for Lampshades, Pattern Design for Magnetic Device. Zheng Di and Fang Binchao wrote two prefaces and indicated that "it is edited according to the standards of pattern design course of new course" on the first page, which shall be taken as teaching material of design

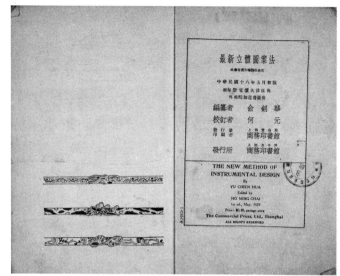

图 3-19　俞剑华编《最新立体图案法》,上海：商务印书馆，1929 年

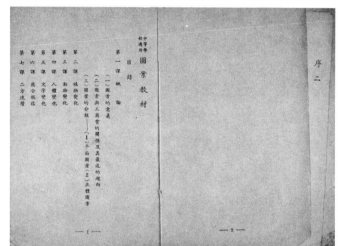

图 3-20　郑棣、方炳潮撰编《图案教材》,浙江正楷印书局，1936 年初版，目录

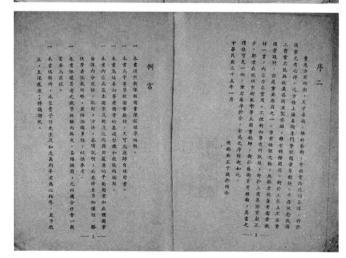

图 3-21　郑棣、方炳潮撰《图案教材》,浙江正楷印书局，1936 年初版，例言

for secondary school and no-teacher self-study book. The characteristics of this textbook are practical and concise. This textbook has 76 pages in all.

第五部：《基本图案学》

《基本图案学》（职业学校教科书）（图3-22），傅抱石编译，商务印书馆，民国25年（1936年）2月初版。这是一部体系完备的教科书，内容包括图案之体系（图案之意义、图案之必要者），要素与资料，写生与便化（原书使用"便化"），美的感觉，构成形式之原理与法则，要素配列上之调和法，单独模样，二方连续模样，唐草模样构成法，四方连续模样，统觉与错觉，立体美之要件，成形法，器体面之装饰。

The Fifth Textbook: *Basic Pattern Studies*

Basic Pattern Studies (Illustration 3-22) translated and compiled by Fu Baoshi was firstly published by the Commercial Press of Shanghai in February 1936. It was a systematic and contextual textbook for vocational schools which could illustrate the System of Pattern Design (The Meaning of Pattern Design, the Necessity of Pattern Design), Elements and Materials, Sketch and Change, Aesthetic Feelings, Guidelines and Principles of Composition Forms, Reconcile Methods of Element Configuration and Arrangements, Singular Matrix Two-Dimensional Successional Matrix, Matrix of Cursive Style in the Tang Dynasty, Four-Dimensional Matrix, Apperception and Misconception, Main Elements of Three-Dimensional Aesthetics, Methods of Modeling and Decorations of Container Facets.

图3-22 傅抱石编译《基本图案学》（职业学校教科书），民国25年（1936年）2月初版

第 3 章　中国近代设计文化概论

第六部:《基本工艺图案法》

The Sixth Textbook: *Methods of Basic Crafts and Patterns*

　　《基本工艺图案法》(图 3-23~图 3-25),傅抱石编译,商务印书馆,民国 28 年(1939 年)3 月初版。此书分为绪论,器体之组成(线之用法、自基本形之器形组成、器形组成余论),器体之装饰(装饰总论、面之分割及色彩之装饰、全装饰面之装饰、散点状装饰、装饰面一部分之装饰、附加工作及附加物之装饰)三大部分。书前的凡例写道:"工艺美术,原以国家民族及时代而各各不同。吾人接触某国工艺品,即可从纹样色彩窥其人民之趣味及其文化之高下。"①书中又提出:"图案之原则即美之原则也。"可见,图案设计凝练了造型、式样、色彩、装饰等美的要素,图案设计也反映了人们的文化特质。

　　Methods of Basic Crafts and Patterns（Illustration 3-23 to Illustration 3-25）was translated and compiled by Fu Baoshi. The first edition of the book was published by the Commercial Press of Shanghai in March 1939. It was composed of three major sections including Introduction, Container Composition（Linear Methods, Basic Container Composition and Container Composition Reflections）and Container Decoration（General Decoration Theory, Facet Division, Colorful Decoration, Overall Decoration, Scattering Decoration, Facet Decoration, and Decorations of Additional Parts）. The explanatory notes of the book wrote as follows: arts and crafts would be different because of different nations and ages. When we came into contact with some countries's arts and crafts, we could glimpse their people's interests and cultures according to pattern colors.①

图 3-23　傅抱石编译《基本工艺图案法》,商务印书馆,1939 年 3 月初版,封面(左)

图 3-24　傅抱石编译《基本工艺图案法》,商务印书馆,1939 年 3 月初版,凡例(右)

---

① 傅抱石编译,《基本工艺图案法》,商务印书馆,1939 年 3 月初版,第 1 页。(Fu Baoshi.（1939）*Methods of Basic Crafts and Pattrens*.Commercial Press of Shanghai.P1）

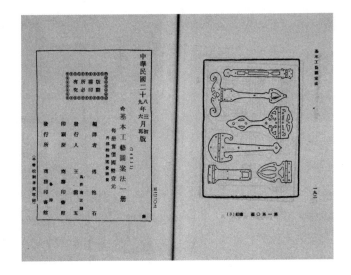

图 3-25 傅抱石编译《基本工艺图案法》，商务印书馆，1939 年 3 月初版，封底

As mentioned in the book, "the principle of pattern design is the principle of aesthetics." This obviously showed that pattern design actually condenses the elements of aesthetics including modeling, style, color and decoration.

第七部：《普通平面图案画法》

The Seventh Textbook: *Techniques of General Graphic Pattern*

《普通平面图案画法》（图 3-26、图 3-27），茅剑青编，上海北新书局，1936 年初版。全书 82 页，书前彩色图案范图三幅，编辑大意、江问渔潘仰尧序、编者自序各一篇，共八编，分别是概论、原则、组织法、取材、写生及意化法、画法、用具、配色。此书图解与文字相得益彰，并附参考表格，甚为实用。

Techniques of General Graphic Pattern(Illustration 3-26, Illustration 3-27), was written by Mao Jianqing, totaling 82 pages, with three colored pattern models before text, including General Idea of Editing, Preface of Jiang Wen Yu Pan Yang Yao, Preface of Editor, totaling eight volumes, including Introduction, Principles, Organization Methods, Draw Materials, Sketching and Imagery Methods, Techniques of Pattern Design, Tools, and Color Matching. The diagram and text of this book brought out the best in each other, attached with form for reference, which could be quite practical. This book was published Shanghai Beixin Press in the beginning of 1936.

第 3 章　中国近代设计文化概论

Chapter Three: Topic on Chinese Design Culture in Modern Times

图 3-26　茅剑青编《普通平面图案画法》，上海：北新书局，1936 年，封面

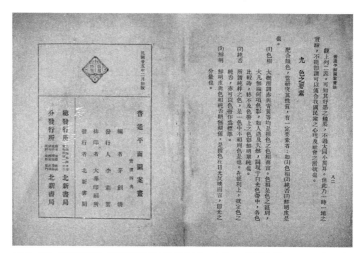

图 3-27　茅剑青编《普通平面图案画法》，上海：北新书局，1936 年，版权页

第八部：《实用图案画法》

The Eighth Textbook: *Technique of Practical Pattern Design*

《实用图案画法》（图 3-28、图 3-29），陆旋编著，上海新民图书馆民国 10 年（1921 年）8 月初版。全书讲述理论与技法部分 104 页，后附 100 幅实用图案画法的范图，除绪言和结论外共八章，分别是总论、色彩、图案原则、描法论、图案资料、自然与变化、纹样形式与感应、平面模样组织法；附录图案净写手续一篇。此书的优势在于：编写条理清晰、文言文简练概要，为师范工业图案教学的适合教材。

Technique of Practical Pattern Design(Illustration 3-28, Illustration 3-29) written by Lu Xuan was firstly published by Shanghai Xinmin Library in August 1921. The book mainly represented the theories and techniques with 104 pages, attached 100 legends of practical pattern design techniques. Apart from Preface and Conclusion, the main body of the book was divided into 8 chapters, including Overview, Color, Pattern Principle, Talks on Tracing Methods, Pattern References, Nature and Variations, Pattern Forms and Interaction, and Graphic Pattern Organization, Appended the Procedure of Pattern Design Sketching. The advantages of this book are that, it has a well-organized compilation, and the classical style of Chinese writing is quite concise and summary, which could be used as teaching materials for normal industrial pattern design education.

图 3-28 陆旋编《实用图案画法》，上海：新民图书馆，1921 年，封面

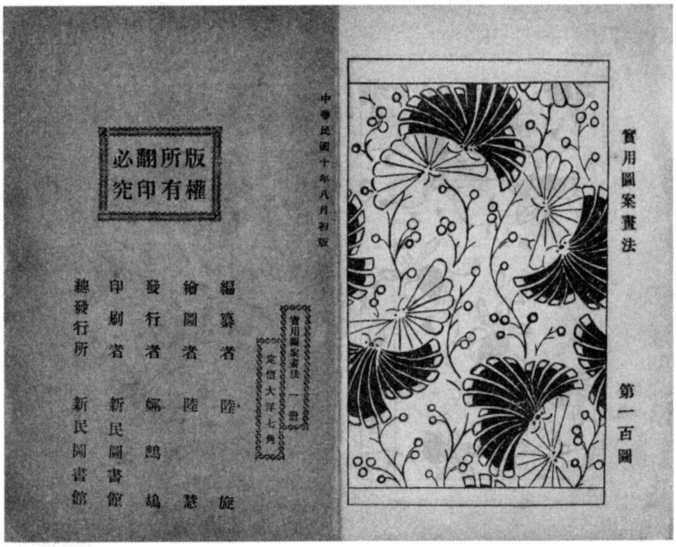

图 3-29 陆旋编《实用图案画法》，上海：新民图书馆，1921 年，版权页

第九部：《图案画法》

《图案画法》（图 3-30、图 3-31），编者朱西一，上海中华书局，民国 36 年（1947 年）初版。全书 64 页，共四章，分别是总说（图案的起源、需要和种类），资料和变化，平面图案画法，立体图案画法。这部教材结尾写道："要把图案应用到食、衣、住、行各方面去，以美化人类的生活，发扬人类的文化。"[①]全书插图丰富、图文并茂、解析简明扼要。

The Ninth Textbook：*Technique of Pattern Design*

Technique of Pattern Design（Illustration 3-30, Illustration 3-31), written by Zhu Xiyi was published by Zhonghua Book Press of Shanghai in 1947. The textbook could consist of four chapters with 64 pages, including Overview, Patterns' Origin, Needs and Types, References and Variations, Techniques of Graphic Pattern, Techniques of cubic pattern. The textbook concluded that in ordern to beautify human life and promote human culture, it would be necessary to apply patterns to all directions such as food, clothing, shelter and transportation. The textbook is full of excellent illustrations and corresponding texts, as well as compendious analyses.

---

① 朱西一主编《图案画法》，上海中华书局初版，1947 年，第 64 页。（Zhu Xiyi. (1947) *Technique of Pattern Design*. Shanghai: Zhonghua Book Press. P64）

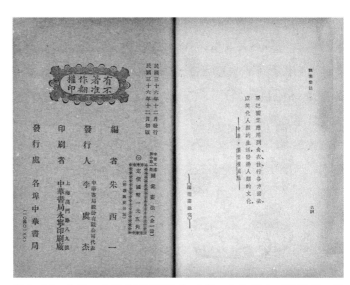

图 3-30 朱西一主编《图案画法》,上海：中华书局,1947 年,封面

图 3-31 朱西一主编《图案画法》,上海：中华书局,1947 年,版权页

第十部：《现代工艺图案构成法》

The Tenth Textbook : *Composition Methods of Modern Technological Patterns*

《现代工艺图案构成法》(1–3 编)（图 3-32~ 图 3-34）此书为现代应用美术专科教本，原著楼子尘，上海形象艺术社 1933 年初版。全书分三编，第一编为图案总说，分别是图案的目的、意义、形式原理、形式法则、资料及变化法六章，附图例 17 幅；第二编为平面图案，分别是平面图案的意义、平面图案的分类、单独模样、边缘模样、四方连续的意义、四方连续的分类、散点模样、不规则的散点模样、连缀模样九章，附图例 50 幅；第三编为立体图案，分别是立体图案的意义、立体模样基本形之构成、立体模样分部和纵横的比较、立体模样

The book of Composition Methods of Modern Technological Patterns（1-3）(Illustration 3-32 to Illustration3-34) was used as teaching material for modern applied fine art, originally written by Lou Jiechen consisting of three parts, and firstly published by Shanghai Image Art Club in 1933. The first part was divided into six chapters, including Overview of Pattern Design, Pattern Design Purposes, Significance, Forms and Principles, Forms and Models, and References and Variations, appended seventeen legends. This second parts was about graphic pattern design, divided into nine chapters, including Significance of Graphic Pattern, Categories of Graphic Pattern, Single Pattern, Edge Pattern, Significance of Consecutive Four-Square Pattern, Categories of Consecutive Four-Square Pattern, Scattering Pattern, Irregular Scattering Pattern,

的实用和适合、立体模样表面的装饰、各部的装饰和适合的模样、模样的描法七章，附图例43幅；每编（章）先论述理论，再补充插图，论述详细，书前自序中提及的"在给生产教育新兴所要求图案科的构成方法及状态，整理一下，俾教授图案科教育家有一定的系统。"为编此书之目的。

诚然，随着"西学东渐"思潮，西方的图案理论与实践及其教学体系传播到中国，或者从当代对设计的解析来看，各式各样的图案书籍在中国编译出版，使西方的设计思想与经验在中国呈现出来并运用于设计实践之中。

and Cluster Pattern, appended fifty legends. the third part was about cubic pattern, divided into seven chapters, including Significance of Cubic Pattern, Composition of Basic Cubic Pattern, Comparison of Segments, Vertical and Horizon of Cubic Pattern, Application And Fitness of Cubic Pattern, Surface Decoration of Cubic Pattern, Decoration of Different Segment and Suited Pattern, Tracing Methods of Pattern, appended forty-three legends. Each part (chapter) had its discourse theory firstly and appended illustrations, with detailed descriptions. Author's preface: The purpose of writing this book was "to sort out the constructive method and situation of pattern design for the emergence of productive education, as well as to provide a standard system for educators in teaching pattern design."

Indeed, as the ideological trend of Western Learning, the Western pattern design theory, practice and teaching system spread to China, or judging by the contemporary on the interpretation of design, all kinds of pattern design books published in China, which would make the Western design ideas and experience present in China and apply to design practice.

图 3-32 楼子尘原著《现代工艺图案构成法》，上海形象艺术社，1933年，封面

第 3 章 中国近代设计文化概论

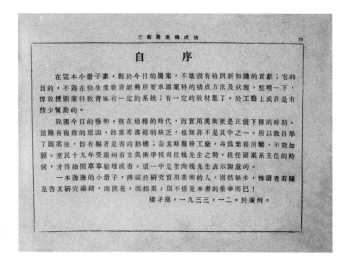

图 3-33 楼子尘原著《现代工艺图案构成法》，
上海形象艺术社，1933 年，自序

图 3-34 楼子尘原著《现代工艺图案构成法》，
上海形象艺术社，1933 年，版权页

3.3 引鉴：西方文化影响下的中国设计
Import: Chinese Design under the Influence of Western Culture

  1918年，蔡元培先生在北京国立美术专门学校成立时所作的《在中国第一国立美术学校开学式之演说》专题演讲中指出："惟绘画发达以后，图案仍与平行之发展。故兹校因经费不敷之故，而先设二科，所设者为绘画与图案甚合也。"20世纪初期，美术教育的改革家们已意识到西方设计具有可鉴之处，积极传播西方设计理念融入教学实践之中。可见，图案教学最早开启了我国近代设计教育的序幕，奠定了近现代绘画与设计并行发展的基础，20世纪初期我国的绘画与设计教育是同步发展的。蔡元培先生倡导的图案教学理念一方面是中国传统图案形态的传承，一方面引荐西洋的图案形质与美学原理，并且影响了当时一批美术教育家以及从事图案设计的人士。

  In 1918, Cai Yuanpei did his keynote speech on "Opening Speech of the First National Art School of China" for the establishment of the Beijing National Art School. He indicated that "only after painting becomes developed can patterns design implement parallel development, and that's why this school set up two disciplines at first because of the insufficient expenditure, and the result was even the integration of painting and pattern design." Reformers of art education were realized that they could learn a lot from Western design. They were spreading Western design concepts and integrating it into teaching practices actively in the early 20th Century. Thus, it can be seen that pattern design is the first to start the prelude of modern design education in China, which lays a foundation of the parallel development of modern and contemporary painting and design. Thus, painting and design education were implemented synchronous development in China in the early 20th century. On the one hand, Mr. Cai Yuanpei's philosophy of pattern art teaching inherited the form of Chinese traditional pattern styles; on the other hand, it introduced the nature of pattern design and the principles of

随着 20 世纪初期出版业的技术革新，各种具有代表意义、体现设计方法与内容的图案教材以及各类出版物相继问世，成为我国早期传播西方设计教育、设计理论、设计方法与内容的史料。这些引荐的图案教材所反映的特点与西方艺术教育体系的传入密切关联，借鉴西方、取长补短，对中国近现代设计教育的发展具有启智作用。顺应当时的新风尚，视觉文化呈现丰富多彩的局面。清光绪二十二年（1896 年）已出现展现中西方设计文化的广告画作品，诸如，在上海发行的商业海报《沪景开彩图》（图 3-35），该作品被认为是中国最早的一件月份牌广告画作品。随后，大量反映此时人们审美文化的标志设计、书籍装帧、月份牌、广告设计、橱窗设计、画报、期刊等陆续涌现，从而形成这个时期特有的中国设计风格。

笔者认为，文化源于历史发展，也源于人们的生活。任何一种文化现象日积月累，一旦形成某种习惯扎根稳固便会成为具有共识的文化传统或思想内核。回顾 20 世纪初期中国文化的发展，

Western aesthetics, and exerted certain influence on a number of art educators and those engaged in pattern design.

With the technological innovation of the publishing industry in the early 20th century, a variety of pattern design textbooks and publications implicating representative significance, reflecting design methods and contents came out one after another and became the early diffusion of historical materials of Western design education, design theory, design methods and contents. We used design experiences of Western countries for reference and drew on each other's merits. These recommended pattern textbooks mirrored the characteristics had close relationship with introduction of Western art education system and showed special effects and inspiration for the development of modern design education in China. Conform to the new tendency at that time, the visual culture presented various prospects. Advertising works which expressed Chinese and Western design cultures already to be appeared. For example, the commercial poster "Colorful Scene in Shanghai (Hu Jing Kai Cai Tu)" (Illustration 3-35) published in 1896 (the twenty-two-year in Qing Dynasty). This commercial poster ought to be considered to be one of the earliest design works of calendar advertising posters. Soon afterwards, numbers of logo design, book design and bindings, calendars, advertising design, shop-window display, pictorials and journals that could reflect the aesthetic cultures of people at that time were springing up. Thereby, the unique Chinese design style of this period had been formed.

The author thinks that culture comes from historical development and people's life. Any kind of cultural phenomenon accumulates over a long period. Once cultural phenomenons were formed solid rooted habits, they will become a consensus of cultural traditions or ideological core. Reviewing the early

总体呈现"洋为中用、兼收并蓄"的特点，设计文化是其中不可忽视的一个亮点，格外鲜明地突出两个关键词，即"新"和"现代"。在很多情形下，两者互训通用，诸如：当时的出版物、杂志与报刊中提出"新工艺"代表"现代工艺"，"新文学"寓意"现代文学"，"新绘画"也指代"现代绘画"，读者能够领会其含义是相同的。此时，在"新文化运动"影响下，人们的思想、观念、生活、行为各方面涌现一股"革新"的力量，并且民族文化与物质文化相融合，产生各种特色鲜明、图案新颖、别具一格的设计实物，仿佛营造了一个推陈出新的"设计新时代"。

20th century, the Chinese caltural development was appeared the characteristics of "adapting foreign things for Chinese use Combining Western and Eastern Classical culture". Design culture is a bright spot which cannot be ignored, particularly highlighted two key words, the "new" and "modern". In many cases, these two words are both common and mutual commenting, the word of "new" is equivalent to the word of "modern". For example, at that time, publications, magazines and newspapers advanced "new craft" referred to "modern craft", "new literature" referred to "modern literature", and "new art" referred to "modern art", readers could understand that they ought to have the same meaning. At this point, under the "New Art Movement", aspects of people's thoughts, ideas, life, behavior emerged a force for "innovation", and with the integration of national culture and material culture, produced all kinds of characteristic, novel design, unique physical design objects, as if building a new "new design age".

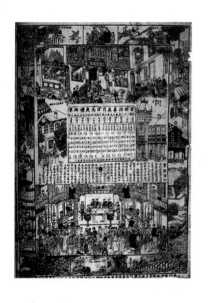

图 3-35　上海发行的商业海报《沪景开彩图》，1896 年，中国最早的广告画

思考题：

1. 请说出你印象深刻的中国早期设计教育家及其代表作。
2. 请采用双语的方式，谈谈我国近代设计的创新思想。
3. 结合范文，请用英语表述一部中国近代设计教材的主要内容。

第 4 章 西方古典时代至 18 世纪设计文化选读

Chapter Four: Selected Readings of Western Design Culture from Classical Era to 18th Century

[本章导读]

西方古典时期处于孕育文化的早期阶段,这一时期可考察的古典文化实例很有限。古希腊的雅典神庙、古罗马的万神庙是西方古典时代的文化象征。传承至今的一部反映西方文化精髓的代表作《建筑十书》,其原著者:维特鲁威 Marcus Vitruvius Pollio,公元前 1 世纪古罗马的御用工程师、建筑师,他的这部论著西方最早提出建筑设计的三要素"实用、坚固、美观"的著作,对现代设计产生深远的影响;此著作最早总结出人体结构的比例规律,可谓现代"人机工程学"的起源。

西方中世纪时期也是暴君专制的时期,但作为这个时期设计成就的探寻,主要体现在 12 世纪初期,在法国卡佩王朝(Capetian Dynasty)领导下,法国人热衷于建造具有"天堂般光辉"的建筑,于是,诞生了展现新兴建筑文化的"哥特式"建筑。西方文艺复兴时期,从共和意识、外交历史、自然科学、人文主义、文化动态乃至艺术发展各个领域的复兴,象征着"古典文化再生"。虽不能一一列举,但笔者选取了与设计文化息息相关的实例。西方巴洛克时期、洛可可时期,分别代表着两种"对比"强烈的设计文化,此时,中国传统纹样设计、园林设计以及瓷器装饰在西方备受关注与青睐,可谓流行于西方的"中西合璧"设计时尚。

本章精选古典时期、中世纪时期、文艺复兴时期、巴洛克时期、洛可可时期比较典型而突出的西方文化史料,特别是与设计学领域息息相关的内容,采用双语对照的方式呈现,希冀为广大读者勾勒出一个西方早期设计文化的轮廓图。

4.1 西方古典时代的设计文化
Design Culture of Western Classical Era

"希腊雅典神庙"是西方古典建筑的代表之一,也是西方古典时代的文化象征。当时这座建筑采用了镀金、彩绘、雕刻等手法,其原貌呈现色彩绚丽、金碧辉煌的华美壮观景象;建筑师运用了西方最古老、最典型的两种柱式,即陶立克柱式(Doric order)与爱奥尼柱式(Ionic order),讲究均衡与比例之美;古希腊人崇尚"艺术想象"和"追求完美"的文化意念,整座建筑体现了"近乎完美"的人体尺度美感(图4-1、图4-2)。

Temple of Athens is not only one of the representatives of the Western Classical architecture, but also a cultural symbol of Western Classical Era. At that time, this architecture adopted gold-plating, colored drawings, and carving techniques and so on and the original architectural appearance showed colorful, glorious, and magnificent scene. Architects applied the oldest and the most typical two types of Western columns, the Doric order and the Ionic order, and strived for the beauty of balance and proportion. The ancient Greeks advocated the cultural thoughts of artistic imagination and the pursuit of perfect. The whole building could reflect the "approximately perfect" human scale aesthetics(Illustration 4-1, Illustration 4-2).

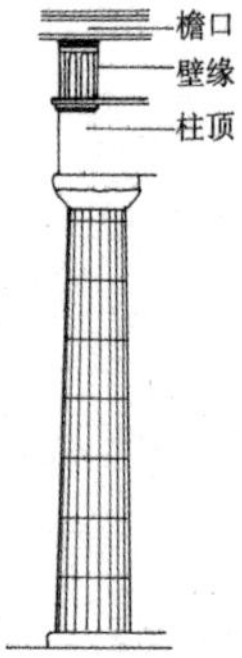

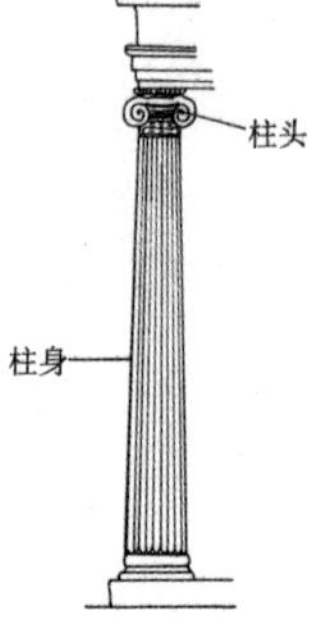

图 4-1 陶立克柱式(Doric order)
(手绘表达:应宜文)(左)

图 4-2 爱奥尼柱式(Ionic order)
(手绘表达:应宜文)(右)

古罗马的"万神庙"建筑留给人们庄严、圣洁、沉静的印象,圆形大厅的顶部中间有一个直径为8.2米的圆孔,这种建筑造型可以追溯到古罗马的文化渊源,罗马人认为"圆顶是天空的象征",这座建筑正中的圆孔代表了天神与地上人们的通灵之处,建筑必须符合程式化的礼仪规范。此外,这座建筑设计恰如其分地体现了罗马人崇尚得体(decorum)、美观(venustas)、实用(utilitas)的建筑文化观念(图4-3、图4-4)。

Pantheon of ancient Rome architecture brings people the impression of solemn, holy and quiet. There is a hole with diameter of 8.2 meters at the top middle circular hall. This architectural style can be traced back to the culture of ancient Rome as the Romans believed the "dome is a symbol of heaven". The median circular hole of the building represents the psychic place of people and God. That means construction must comply with the stylized norms of etiquette. Furthermore, this architecture design is just perfect to embody the Rome's architectural culture concepts of appropriation (decorum), pleasing to the eye (venustas), and pragmatism (utilitas) (Illustration 4-3, Illustration 4-4).

古典时期的设计文化在陶器、玻璃制造器皿以及家具设计领域也表现丰富,诸如意大利阿雷佐(Arezzo)生产的阿雷陶器(Arretine Ware)、默勒石玻璃器皿(Murrhine Glass,又被称为"亚宝石器皿")、波特兰花瓶(Portland Vase)、克里斯莫斯椅子(Klismos)等,并且在设计制造上汲取欧洲神话和古典文学中的各种"神"的形象,反映了严谨而精美的古典文化气质(图4-5~图4-7)。

During the classical era, design culture in the field of pottery, glass utensils and furniture design showed abundant, such as the Arretine Ware made in Arezzo Italian, the Murrhine Glass, the Portland Vase, and Klismos Chair. These design and manufacture absorbed a variety of "god" images from European mythology and classical literatures and reflected the rigorous and exquisite classical culture temperament (Illustration 4-5 to Illustration 4-7).

图4-3 古罗马万神庙建筑正门
(手绘写生:应宜文)(左)

图4-4 古罗马万神庙建筑室内圆顶(右)

图 4-5 意大利阿雷佐（Arezzo）生产的阿雷陶瓷（Arretine Ware）

图 4-6 波特兰花瓶（Portland Vase）
（手绘表达：应宜文）

图 4-7 克里斯莫斯椅子（Klismos）
（手绘表达：应宜文）

值得一提的是，维特鲁威（Marcus Vitruvius Pollio，约公元前 180 年~前 15 年之后）的《建筑十书》（De Architectura）是西方从古典时代保存至今唯一的一部关于建筑、绘画和雕塑的专著，也是西方传承至今的一部反映古代设计思想的专著。这部专著以建筑师为中心，还论及绘图、历史学、哲学、音乐学等诸多领域，蕴含着西方文明的精髓。这本著作的成书年代大致在公元前 32~前 22 年间，1486 年罗马第一次出现这部著作的印刷版形式。

It is worth mentioning that, Marcus Vitruvius Pollio's *De Architectura* is the only work related to architecture, painting and sculpture monograph which have been preserved from the classical era up to now. It is also the ancient design philosophy monograph inherited so far. This monograph describes architect designer as the core, and explains drawing, history, philosophy, music and many other fields, which would contain the essence of Western civilization. This work was written between BC 32- BC 22 years. The first edition published in the form of printing appeared in Rome in 1486.

4.2 西方中世纪的设计文化
Design Culture of Western Middle Ages

兴盛于中世纪的日耳曼民族不再崇尚古典主义文化，而是显示一种与其文化同一的充满"运动感"的设计形式，受到这种充满野性、反复剧烈的文化习俗影响，当时的装饰物件中出现各种"重复的构图"、"活力的线形"等装饰风格。这个时期的金属工艺的设计，诸如铅玻璃上色、切割宝石镶嵌以及金属模铸物件等，往往呈现错综复杂的图案设计，并以小型工艺品为主。

12世纪初期，在法国卡佩王朝（Capetian Dynasty）领导下激发了百姓的民族自豪感，激情澎湃的法国人热衷于建造具有"天堂般光辉"的建筑，于是，一种展现新兴建筑文化的"哥特式"建筑诞生了。这种建筑的室内环境采用彩色玻璃窗的墙体，建筑立面雕刻精美，采用尖拱（ogival arch, pointed arch）和肋架拱顶（ribbed vault）

The Nordic Nation flouring in Middle Ages no longer advocated the classical culture. Adversely it fully showed "a sense of movement" which would be aligning to their culture. Being affected by this wild, and violent repeated cultural custom, the decorative objects appeared various "repeat composition" and "dynamic alignment" and other decorative styles. During this period metal technology design, such as lead glass coloring, cutting precious stones inlaid and metal casting objects, etc., tended to present a kind of complicated pattern design, which would mainly apply to small crafts.

In early 12th century, the sense of national pride was inspired in common people under the leadership of French Capetian Dynasty. The Passionate French were enthusiastic about the construction of "heavenly glory" buildings; therefore, the "Gothic" new-style architectural culture was born. The indoor environment of the building adopted wall made of stained glass windows, the building elevation was exquisitely carved. The basic building components such as ogival arch and ribbed vault were used, thus giving people

为基本建筑构件，能给人们一种向上升腾的视觉效果。诸如：博韦大教堂（Beauvais Cathedral）、亚眠大教堂（Amiens Cathedral）、兰斯大教堂（Reims Cathedral）等（图4-8、图4-9）。

因受当时建筑文化的影响，在室内设计、家具设计、器皿设计、服装设计等各个领域新生一种以"高耸趣味"为主导的时尚文化。诸如：当时，无论男士、女士的服装都以"尖锐悠长"为审美典范；普遍流行一种"尖头型"鞋子设计来体现哥特式精神；能工巧匠们更是把家具打造得好像一座微缩版的哥特式建筑（图4-10）。

与此同时，宗教文化和宫廷文化也体现在当时设计领域之中。家具上雕刻有圣徒形象，尤其皇室的家具设计，采用象牙和黄金制成，工艺精湛、材质精良，追求豪华壮观的效果。服饰上采用鹫、狮子、孔雀等纹样，符合伦理和礼仪规范，追求金碧辉煌、至高无上的效果，以显示一种"荣耀感"。

a visual effect of ascension, for example, the Beauvais Cathedral, the Amiens Cathedral and the Reims Cathedral (Illustration 4-8, Illustration 4-9).

Affected by the architectural culture then, a towering-dominated new fashion culture aroused in fields including interior design, furniture design, vessels design and clothing design. For example, at that time, clothing for both men and women took "sharp and slim" as the aesthetic paradigm; a type of "sharp end" shoes design was prevalent and reflected the Gothic spirit; furthermore, the skilled craftsmen built furniture into a seeming miniature the Gothic Style building (Illustration 4-10).

At the same time, religious culture and royal culture were embodied in the design fields then. The images of saints were carved on the furniture. In particular, the royal furniture design was made from ivory and gold materials, the workmanship was exquisite, the materials were excellent and a kind of grand, spectacular effect was pursued. The clothing used patterns such as vulture, lion and peacock and so on, which was in line with the ethics and etiquette standards and sought a type of splendid, magnificent and sovereign effect, so as to display a sense of "honor".

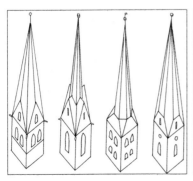

图4-8 哥特式建筑顶部尖拱结构（ogival arch）
（手绘表达：应宜文）（左）
图4-9 欧洲哥特式建筑写生
（手绘表达：应宜文）（中）
图4-10 受哥特式建筑文化影响下"尖锐修长"风格的时尚服装设计
（手绘表达：应宜文）（右）

4.3 西方文艺复兴的设计文化
Design Culture of Western Renaissance

欧洲文艺复兴发端于14世纪的意大利，15世纪中后期在欧洲各国盛行，16世纪达到鼎盛。文艺复兴不仅仅指恢复古希腊、古罗马的艺术，更是代表着"古典文化再生"，它包含思想、文学、艺术在内的所有文化领域。这场复兴运动以"复古"为启发，以"发现"传统文化为根基，以实行再创新为主旨，对当时及其后两三百年的西方设计产生了深刻的影响。

首先，在建筑领域文艺复兴时期的建筑师善于将古典优雅之美与中世纪哥特式建筑风格结合一体，较为典型的例子是布鲁内莱斯基（Filippo Brunelleschi，1377—1446）设计的佛罗伦萨大教堂圆顶。阿尔伯蒂（Leon Battista Alberti，1404—1472）撰写了《论建筑》一书提出"美"的三个标准，即数字（numbers）、比例（finitio）、

European Renaissance originated from Italy in the 14th century, prevailed in European countries in the mid and late 15th century and culminated in the 16th century. The Renaissance not only refers to the restoration of ancient Greek and Roman arts but also represents "regeneration of classical culture", covering all cultural fields of ideology, literature and art. The Renaissance movement, enlightened by "restore ancient ways" and rooted in "discovering" traditional cultures with the purport of re-innovation, had profound impacts on Western design then and the following two to three hundred years.

Firstly, in the architecture field, architects during the Renaissance period were adept at integrating the elegant beauty of classics with the Gothic style of the Middle Ages. A typical case was the dome of the Florence Cathedral which would be designed by Filippo Brunelleschi (1377-1446). Leon Battista Alberti (1404-1472) wrote the book titled *On Building*, put forward three standards of beauty such as numerals (numbers), proportion (finitio), and distribution (collocatio) and explained the problems of

图 4-11　文艺复兴时期的欧洲圆顶建筑写生之一
（手绘表达：应宜文）

图 4-12　文艺复兴时期的欧洲圆顶建筑写生之二
（手绘表达：应宜文）

分布（collocatio），阐释了审美与装饰的问题。这部建筑学专著也是在维特鲁威的《建筑十书》理论基础上再提出新的见解（图 4-11、图 4-12）。

其次，在手工艺领域文艺复兴时期的陶瓷工艺最早在意大利繁荣起来。当时技术工匠采用人物与花卉植物配合的纹样，还有一些描绘精美颇具文化气息的图样。金属工艺在当时是艺术家、雕塑家必须学习的一门手艺，诸如佛罗伦萨洗礼堂的青铜门饰是雕刻家基伯尔提的作品。意大利、法国、德国的工艺设计摆脱了中世纪的设计风格，出现了将金属材质与珐琅、玉石、陶瓷、象牙等不同材质互相组合的设计创新之举（图 4-13、图 4-14）。

aesthetics and decoration. The monograph of architecture also advanced new views and concepts based on the theory of Vitruvius's *De Architectura*.（Illustration 4-11, Illustration 4-12）.

Secondly, ceramics in the Western Renaissance prospered in Italy for the first time in the handicraft field. At that time, crafters used figures and flowers as the patterns combination as well as some designs with exquisite descriptions and cultural atmosphere. Metal technology was a technique that artists and sculptors must learn back then. For example, the bronze door decoration of Florence auditorium （Battistero di San Giovanni）was the artistic work of sculptor Lorenzo Ghiberti. Craft design of Italy, Singapore and Germany broke away from the Middle Ages design style in terms of technology, and made the design innovative move of combining metal with different materials such as enamel, jade, porcelain, ceramics and ivory（Illustration 4-13, Illustration 4-14）.

文艺复兴时期的家具设计体现浓郁的西欧文化风俗。15世纪初期，意大利婚礼家具中必有"卡索奈长箱"，这种家具成对制作，两个长箱上分别刻有新郎、新娘的家徽，家具上的装饰也必请设计家、雕刻家、画家来完成，因此，装饰题材有古代神话、丘比特、几何图样、花卉果实等皆体现出人们淳朴的美好愿望。座椅家具更有传统文化印迹，意大利著名诗人但丁早年曾当选佛罗伦萨共和国行政官，他企盼回到故乡，历经千辛万苦，为后人留下一部经典史诗《神曲》。根据他执政官时的坐席设计师发明了"但丁椅"，即左右四根"S"形粗腿，采用雕刻和镶嵌工艺的外部装饰，至今，这款"但丁椅"在欧洲的公众礼仪活动中仍被广泛运用（图4-15）。

Renaissance furniture design reflected the rich Western European cultural custom. In early 15th century, there must be the Cassone Long Boxes among all furniture in an Italian wedding. The long boxes were made in pairs, with the family heraldries of the bride and groom carved on each one of them, and the furniture decoration was completed by some designer, sculptor and painter, therefore, the decorative themes covered ancient myths, Cupid, geometric patterns and flowers and fruits, all of which would gave expression to the unsophisticated good wishes of people. The furniture seat could carry more traditional cultural trace. The famous Italian poet Dante was elected administrator of the Republic of Florence in his early years, later, he longed to return to his native land after untold hardships and left behind a legacy of a volume of classic epic "the Divine Comedy". According to his seat in office, the designer created the "Dante Chair", which would have four S-shaped thick legs and adopt exterior decoration of carving and inlay. Up to now, this type of the "Dante Chair" have been widely used in public European ceremonies.

图4-13 佛罗伦萨洗礼堂的青铜门

（实地拍摄：王颖）

图4-14 佛罗伦萨洗礼堂的青铜门装饰细节图

（实地拍摄：王颖）

图4-15 纪念意大利著名诗人但丁而设计的"但丁椅"，这款椅子造型优雅均衡在当时很受欢迎

4.4 巴洛克时期、洛可可时期的设计文化
Design Culture of Baroque and Rococo Period

欧洲人曾把17世纪称为"巴洛克时代"（Baroque Era）。"巴洛克"是一种艺术风格的总称，也代表了一种独特的建筑文化。巴洛克设计的主要成就集中体现在建筑领域，诸如贝尔尼尼（Gian Lorenzo Bernini，1598—1680）设计的梵蒂冈圣彼得教堂的椭圆形广场、法国凡尔赛宫建筑设计等典型建筑。巴洛克的设计风格往往采用多样化的艺术处理与开放性的空间构造，兼有延伸感和凝聚感的空间设计。这种设计风格以奢华复杂、炫耀夸张的曲线形状和螺旋形态作为装饰，给人们以辉煌宏大之感，并主张以规范化与秩序美为核心的设计审美文化。（图4-16）。

西方的视觉文化史上，洛可可的设计以"愉悦、优雅、妩媚"之风格而著称，最早出现在18世纪欧洲的室内设计领域。这种设计风格与文

The Europeans called the 17th century "Baroque era". The "Baroque" was not only a kind of artistic style, but also represented a unique architectural culture. The main achievements of Baroque design were embodied a concentrated reflection in the field of architecture such as typical architectures of the oval square of Vatican St. Peter's Church designed by Gian Lorenzo Bernini (1598-1680), and the architectural design of French Versailles palace and so on. The Baroque style would tend to adopt diversified artistic processing and open space structure, contain spatial design with both extension and cohesion. This kind of design style used some luxury, complexity, exaggerated curve and spiral shapes as decoration elements, and gave people a sense of magnificent grand. It maintained the design aesthetic culture centered on standardization and order beauty (Illustration 4-16).

In the history of Western visual culture, the Rococo was known throughout the world for its pleasure, elegance, and charming and appeared in European interior design field during the 18th century firstly. Such design style was closely

图 4-16　法国凡尔赛宫入口处写生
（手绘表达：应宜文）

化潮流密切相关，当时社交界的女性主义文化占主导地位，因而引领一种柔弱妩媚、浮华精美至极的时尚文化风格。洛可可的设计注重建筑表面的装潢，典型的设计采用涡卷式、交错式的纹样以及中国花鸟人物的各种图样作为装饰。在家具设计方面，洛可可设计追求舒适、简便、造型精巧的设计风格；在室内环境设计方面，典型的洛可可居室设计整体色调清新淡雅，在浅色基础上配以金色、绿色、玫红作为装饰点缀，显示精美、细腻、轻快之特点。

related with cultural trends. Feminism culture dominated in the social circles at that time, thus leading a fashion culture style that's characterized by delicate, charming, ostentatious and extremely exquisite appearances. The Rococo design focused on decoration of the architecture surface. Scrolling and interlacing patterns as well as Chinese flowers, birds, and figures were regarded as typical design ornaments. In furniture design, the Rococo design that highlighted comfort, convenience and deft shapes was pursued. In interior environment design, the typical Rococo room design exhibited the overall tone of refreshing and quietly elegant. The light colors were dotted with gold, green and rose-bengal, thus demonstrating exquisite, fine, bright and brisk features.

正是这段时期，欧洲上层社会对中国瓷器十分青睐，欧洲工匠不断探究中国瓷器的制作工艺，在学习中国瓷器制作基础上，相继出现了一种具有中国瓷光泽度的欧洲软质瓷（soft-paste porcelain）和一种具有中国瓷质感的欧洲硬质瓷（hard-paste porcelain）。与此同时，中国古典艺术和文化影响了欧洲的装饰设计，一些欧洲制造的瓷器上绘有中国风格的古典园林、人物、楼阁等纹饰，并且采用中式红、蓝、青、黄等凸显中国特色的色彩组合，出现专门仿制中国青花瓷风格的各种器皿。值得一提的是，英国设计师齐蓬代尔（Thomas Chippendale，1718—1779）曾致力于研究中国园林、家具设计和建筑文化，并结合简练的英式风格。他设计出"中国式大床"，这件家具采用黄色的丝绸帷幔、中国亭榭式顶部造型和格栅式透雕工艺。当时，富有中国传统文化韵味的图案也曾广泛运用在家具椅背、橱柜、书架上，可谓流行于西方的"中西合璧"设计时尚。

It is this time that the European upper classes found huge flavor in Chinese porcelains and china. European craftsmen continued to explore the production processes of Chinese porcelain. On the basis of studying Chinese porcelain production, two kinds of porcelain were successively appeared such as the European soft-paste porcelain possessing the gloss of Chinese porcelain and the European hard-paste porcelain possessing the texture of Chinese porcelain. At the same time, Chines classical arts and cultures affected European decorative design. For example, some European porcelain were painted with Chinese style patterns of classical gardens, figures, pavilions and other ornamentations and used the highlight color combinations such as Chinese red, blue, green and yellow and so on. There was a variety of utensils special imitation of Chinese blue and white porcelain style. It ought to be worth mentioning that the British designer Thomas Chippendale (1718-1779) would do study and devote himself to Chinese gardens, future design, and architectural culture combining with concise British style. He designed the Chinese style bed which used yellow silk bed-curtain, top modeling of Chinese pavilion style, and grille type fretwork. Patterns full of Chinese traditional culture lasting appeals were widely used in furniture such as the back of chairs, sideboards, and bookshelves. It could be popular in the Western fashion design named a combination of Chinese and Western elements.

思考题：
1. 请采用双语的方式，分别列出古典时期、中世纪时期的代表建筑。
2. 请用英语表述西方文艺复兴时期的设计文化实例。
3. 请比较巴洛克时期、洛可可时期的设计风格特征并用英文解说。

第5章 西方19世纪至20世纪设计文化选读

Chapter Five: Selected Readings of Western Design Culture from 19th to 20th Century

[本章导读]

工业社会早期，传统与现代之间的文化碰撞，使设计发挥出巨大的作用，各种日新月异的文化宣传以及设计展览得到了前所未有的发展，造就了一批新兴西方设计师和早期工业家。19世纪随着技术、产品与工艺的革新，出现一个充满活力的设计进步时代，工艺美术运动源于约翰·拉斯金的理论研究，以威廉·莫里斯为代表的先锋设计师将理论转化为设计行动，这次卓有成效的设计革命，成为"新艺术运动"的先声，拉开了现代设计的序幕。随后，文脉相承的"新艺术运动"重点解决了产品形态与艺术文化分裂等一系列设计问题，继续向传统装饰艺术、传统文化与手工艺取经，达到了"青出于蓝而胜于蓝"的境地。从危难中创生机会的美国"芝加哥学派"，倡导"形式服从功能"（Forms follow function）的设计文化理念，该学派的设计师探索西方传统文化与现代设计之间的平衡，采用新材料新技术、实施设计实践，涌现出一大批新生建筑和地标建筑，展现了当时最先进的建筑设计水平。

本章精选工业社会早期、19世纪末到20世纪初期的工艺美术运动、20世纪初期的新艺术运动、芝加哥学派的设计文化理论与实例，并融入笔者的感悟与思考，追溯西方多元设计文化的沿革，探讨近现代设计领域多元化、跨学科发展的新趋向。

5.1 工业社会早期的设计文化
Design Culture of Early Industrial Society

开端于18世纪60年代的英国工业革命（Industrial Revolution）带来了生产技术、文化、经济各方面的一场变革，使整个社会自此迈向机器时代。在传统与现代之间的文化碰撞和创新发展，使西方设计文化领域呈现"百家争鸣"的态势，同时造就了一批新兴设计师和早期工业家。

工业革命前后的设计风格差异显著：一方面，工业革命之前在欧洲大陆以"新古典主义"设计风格为主流，很多设计产品上再度涌现"希腊神话题材"图案以及各种古典式纹样，英国著名的设计师乔塞尔·韦奇伍德（Josiah Wedgwood, 1730—1795）研制了造型简洁、色泽多姿多彩并且富有装饰潜力的"皇后陶瓷"，这种精致美观又实用的陶瓷远销欧洲各地，备受欢迎。韦奇伍德致力于开发仿制古典风格的新型瓷器，1790年他开始

Beginning in the 1760s, the British Industrial Revolution brought a change in all aspects of production technology, culture and economy and the society ever since stepped into the machine age. The cultural collision and innovative development between tradition and modernism, the field of Western design culture presented the trend of "contention of a hundred schools of thought" and achieved a number of emerging designers and early industrialists at the same time.

Before and after the industrial revolution, the design style was significantly different. On the one hand, European mainland worshiped the new classical style as mainstream before the industrial revolution. Many design products emerged patterns of Greek mythology and all kinds of classical patterns again. British well-known designer Josiah Wedgwood (1730-1795) developed and manufactured the Queen's Ware, which would be concise modeling, rich in color, decorative potentiality and enjoy wide popularity. This kind of exquisite, beautiful and practical ware was very popular and exported to all over Europe. Wedgwood devoted himself to develop new porcelain of imitation classical style and was the earliest person to promote and set up fundamental rule of

生产陶制仿古花瓶并成功地将技术革新运用于古典风格的新型陶瓷，他最早提倡并且设立了工艺生产的基本规则。他在文化与技术领域的革新使落后的传统手工艺转化为大规模工业化生产，他对现代工业设计早期发展作出巨大贡献（图 5-1、图 5-2）。

另一方面，工业革命早期带给西方设计领域的改革举世瞩目，新材料、新技术和新结构凝合创生各种新设计，同时带给人们多元化的设计时尚与审美文化。诸如：英国皇家园艺总监帕克斯顿（Joseph Paxton, 1801—1865）设计了博览会"水晶宫"（Crystal Palace）建筑，采用当时的新型建材玻璃和钢铁，展示工业革命早期的新发明成果。另一个实例是查尔斯·巴里（Charles Barry, 1795—1860）与普金（Augustus Welby Northmore Pugin, 1812—1852）两位建筑师合作设计了英国议会大厦，这座建筑

craft production. In 1790, he started to product antique pottery vase and successfully utilized technological innovations to a new type of ceramic in the classical style. His innovation in cultural and technical aspects enabled lag traditional handicraft be turned into large-scale industrial production and made a great contribution to early development of modern industry design (Illustration 5-1, Illustration 5-2).

On the other hand, the design field reforms caused by the industrial revolution in the early period attracted the worldwide attentions, which can be reflected that various new types of materials, technologies and structures generated the diversified designs as well as design fashion and aesthetic culture for people. For example, the Expo architecture of Crystal Palace designed by the British royal gardening design director Joseph Paxton adopted the temporal new types of building glass and steel showing the new inventions in the early Industrial Revolution Time. Another example is the British Houses of Parliament cooperated by two architects, Charles Barry and Augustus Welby Northmore Pugin. The two architectures not only drew lessons from the gothic style

图 5-1 韦奇伍德研制的皇后陶瓷
（摄影：王颖）

图 5-2 韦奇伍德研制生产的仿制古典风格的波特兰花瓶
（邵宏主编，颜勇、黄虹等编著《西方设计：一部为生活制作艺术的历史》，长沙：湖南科学技术出版社，第 154 页）

不仅借鉴了西方中世纪的"哥特式建筑"风格，还在传统设计基础上衍生创新，成为西方传统文化再利用设计的标志（图5-3、图5-4）。

随之而至，家具设计也延续着哥特式风格样式，庞大厚重、局部精美、外观严谨，家具上的图案以古典的植物纹样、涡卷装饰为主导。纺织工业的技术进步，诞生了许许多多新兴的时装图版与服装设计，并促进了室内装饰业的发展，体现出社会生活中的大众文化意识。印刷技术的革新迎来了西方文化界的春天，设计印刷精美的书籍、报纸、杂志的传播越来越普及化，各种设计展览及文化宣传品与日俱增，广告业也得到了前所未有的发展。

of architecture from the west world in the Middle Ages, but also derived the re-creation on the basis of the traditional foundation of design and became a sign of the Western traditional culture mirror design（Illustration 5-3, Illustration 5-4）.

Accordingly, the furniture design also was characterized by huge and decorous, local exquisiteness, rigorous appearance - a Gothic style. And the patterns of furniture were mainly classical plant patterns and scroll decoration. With the technical progress of textile industry, many emerging fashion layouts and costume design came into being, promoting the development of interior decoration industry and reflecting the public cultural awareness in social life. The innovation of printing technology may prosper the Western spring of cultural circles. It was popular for the spread of designing and printing fine books, newspapers and magazines, and the advertising industry experienced an unprecedented development with the daily changing various design exhibitions and promotional cultural materials.

图 5-3　英国皇家园艺总监帕克斯顿（Joseph Paxton）设计的博览会"水晶宫"（Crystal Palace）建筑
（鲁石著《你应该读懂的100处世界建筑》，西安：陕西师范大学出版社，第266页）

图 5-4　查尔斯·巴里（Charles Barry）与普金（Augustus Welby Northmore Pugin）两位建筑师合作设计的英国议会大厦
（鲁石著《你应该读懂的100处世界建筑》，西安：陕西师范大学出版社，第253页）

5.2 19世纪末至20世纪初期西方设计思潮：工艺没事运动

Trend of Western Design Thoughts from the Late 19th Century to the Early 20th Century：Arts & Crafts Movement

一、工艺美术运动的文化渊源

英国工业革命给社会带来了巨大的变化，传统手工工厂被机器工厂所取代，传统手工业被大机器工业所取代，造成英国社会结构和生产关系发生重大改变，从而对社会文化发展产生了深远影响。工业革命的生产直接导致了产品的设计、制作与销售完全分离，这与传统手工业时代作坊主和工匠既是设计者，又是制作者，甚至是销售者和使用者的状况迥然不同。

工艺美术运动始于工业革命，首先在建筑设计领域率先对传统文化提出了挑战，随后在工业设计领域，工业化批量生产带来生产力的提高，却造成产品设计丑陋、过度装饰、产品设计与艺术文化分裂等各种问题，以至在工业革命发生之后的一段时间

I. Cultural Origins of Arts & Crafts Movement

British Industrial Revolution brought great changes to the society and caused significant shifts of British social structure and production relationship. For example, the traditional handicraft factory was replaced by the machine factory. The traditional handicraft industry was replaced by the large-scale machine industry. Thus, it had a profound influence on the development of social culture. Industrial revolution directly resulted to the thorough separation of designs, manufacture and sale in the production process, which would be quite different from the factory owner and craft men. The factory owner and craft men were also the designers, producers or even sellers and users in traditional handicraft stage.

The Arts & Crafts Movement began in the industrial revolution. It firstly presented challenges to traditional culture in the fields of architectural design. Soon afterwards, it developed in the fields of industrial design, the quantity production had improved the productive force, but has also caused all kinds of problems such as the ugly designs, over-decoration and the separation between the product design

里，急需确立新观念唤起人们对传统装饰艺术、传统文化与手工艺的向往。这种回归的呼声日趋强烈，终于导致在工业革命的发源地英国掀起了工艺美术运动。

19世纪末到20世纪初，工艺美术运动可谓是一场设计运动，也是一场西方传统文化的复兴运动。欧洲乃至西方各国的文化渊源存在一些共通性，审美取向上也存在相似的文化观念，因此这场设计运动迅速从英国蔓延到欧洲大陆的法国、意大利、斯堪的纳维亚半岛，甚至盛行于美国，它的影响力广布在建筑设计、家具设计、陶瓷设计、金属工艺、染织品和平面设计等领域。

二、工艺美术运动的设计思想与设计实践

工艺美术运动以追求传统纹样和哥特式风格为特征，旨在提高产品质量，复兴手工艺品的设计传统；倡导"美术和技术结合"的设计思想，反对"纯艺术"或"纯技术"，促使一批美术家从事产品设计，包括艺术家、设计家和建筑师，威廉·莫里斯（William Morris，1834—1896）是一位主导人物。1888年，在威廉·莫里斯的倡导下，创立"工艺美术协会"（The Arts & Crafts Society），将这场设计复

and artistic culture so that a new concepts was needed later the industrial revolution to arouse people's desires and pursuits to the traditional decorative art, traditional culture and handicraft art. These calls for return were so intense day by day that the Arts & Crafts Movement was created in England, the origin of the industrial revolution.

The Arts & Crafts Movement was not only one kind of design movement, but also a renaissance of Western traditional culture during the late 19th century to the early 20th century. The cultural origins of European and Western countries exit some commonality. There are similar cultural concepts for aesthetic orientation. So, the design movement quickly was spreading from England to France, Italy, Scandinavian Peninsula and other European countries, even popular in the United States. Its influence extended in the fields of architecture design, furniture design, ceramic design, mental craft, dyed fabric, and graphic design and so on.

Ⅱ.Design Theory and Design Practice of Arts & Crafts Movement

Arts & Crafts Movement was in pursuit of traditional veins and gothic style as features, aimed at improving the quality of products, revived design tradition for handicrafts. It proposed the design theory of combination of arts and technology against the pure arts or pure technology, impelled a group of artists engaged in product design including artists, designers, and architects. William Morris (1834-1896) was a leading person. On the initiative of William Morris, the Arts & Crafts Society was founded and carried forward the design renaissance movement to the climax.

兴运动推向了高潮。

工艺美术运动的设计理念总体表现在五个方面：第一，在生产制作方面提倡传统手工艺，明确反对机械化的大批量生产；第二，在装饰风格方面反对单纯从古典出发的复兴风格，以自然为师，推崇自然主义、东方装饰风格以及富有东方艺术特点的设计；第三，从西方中世纪的设计文化中获得灵感，以"哥特式"设计风格为主旨，推崇简约温雅、朴素无华并具有良好功能的设计；第四，强调以设计作品来体现诚恳的态度，反对哗众取宠、华而不实的设计导向；第五，设计流程上，威廉·莫里斯及其追随者们强调设计的服务对象，希望产生一种"为人民服务"的艺术，并能够重新振兴工艺美术的民族传统。

关于工艺美术运动的设计实践，首先关注的是1859年威廉·莫里斯在乌普顿的著名建筑"红屋"（Red House），由菲利普·韦伯（Philips Webb，1831—1915）担任建筑平面设计。此屋的设计充分体现了工艺美术运动在建筑设计方面的思想，创立了建筑设计的四条基本原则：第一，在形态、材料和装饰方面，每个室内空间都是结构和面的逻辑派生；第二，每个室内空间都具有与其功能相适应的个性，同时它又犹如一个连接房间大主题的变调；第三，每个室内空间都应如实地展现其结构设计；第四，

Design ideas of Arts & Crafts Movement overall performance in five aspects. The first, it advocates traditional handicrafts in production and clearly opposes mechanized mass production. The second, it denies a revival style from classicism simply in the decorative style, and learns from nature, proposes naturalism and oriental decorative styles as well as design works with rich oriental art features. The third, it is inspired by the design culture from Western Middle Ages with the Gothic design style as the theme, the author respects a design featuring minimalism and elegance, plainness but good functions. The forth, it stresses to experience the sincere attitude via design works, refuses a design-orientation that pleases the public and is like a white elephant. The fifth, in terms of design process, the William Morris and his adherents stress the service objects of design, hoping to produce a sort of art "serving the people", and to revive the national tradition of emerging arts and crafts.

On design practice of Arts & Crafts Movement, the first concern is the William Morris's famous architecture of Red House designed by Philips Webb (1831-1915) for architectural graphic design in Upton in 1859. The design of this house gives full expression to the ideas of Arts & Crafts Movement in architecture design and four fundamental principles of architecture design have been established. The first, in terms of form, material and decoration, every interior space is the logical derivation of structures and surfaces. The second, every interior space possesses its own personality complying with the function, and it is like a modified tone connecting the grand theme of the room. The third, every interior space should faithfully manifest its structural design. The forth, from large space to minor

每个室内从大空间到小细节都必须使用与整体相协调的材料，使其连贯一体。英国学界认为，莫里斯的"红屋"事务所设计真正拉开了工艺美术运动的序幕（图5-5，图5-6）。

理查德·诺曼·肖（Richard Norman Shaw，1831—1912）与莫里斯是同时代人，他在斯特里特工作室遇见威廉·莫里斯之后深受启迪，两人成为志同道合的合作伙伴，追求并拥有共同的文化艺术观念。诺曼·肖自1963年以后主要从事建筑设计，他日后成为英国家喻户晓的设计家之一（图5-7、图5-8）。

details, materials coordinating with the entirety should be adopted, so as to achieve coherence and oneness. It is considered in British academic circle that the Red House Office designed by Morris actually ought to open the prologue of Arts & Crafts Movement (Illustration 5-5, Illustration 5-6).

Richard Norman Shaw (1831-1912) was Morris's contemporary. After he met William Morris in the Street Studio for the first time, he was deeply inspired by Morris. Then the two men became like-minded cooperative partners who pursued and had common culture and artistic ideas. Norman Shaw had been mainly occupied in architectural design since 1963 and later became one of the well-known designers in Britain (Illustration 5-7, Illustration 5-8).

图5-5　1859年威廉·莫里斯在乌普顿的著名建筑"红屋"（Red House），由菲利普·韦伯（Philips Webb）担任建筑平面设计
（鲁石著《你应该读懂的100处世界建筑》，西安：陕西师范大学出版社，第286页）

图5-6　由威廉·莫里斯设计的可以调节椅背的"莫里斯椅"
（邵宏主编，顔勇、黄虹等编著《西方设计：一部为生活制作艺术的历史》，长沙：湖南科学技术出版社，第199页）

第 5 章　西方 19 世纪至 20 世纪设计文化选读

图 5-7　由理查德·诺曼·肖（Richard Norman Shaw）设计的伦敦老天鹅住宅建筑外立面

（邵宏主编，颜勇、黄虹等编著《西方设计：一部为生活制作艺术的历史》，长沙：湖南科学技术出版社，第 203 页）

图 5-8　由理查德·诺曼·肖（Richard Norman Shaw）设计的伦敦老天鹅住宅建筑客厅室内设计

（邵宏主编，颜勇、黄虹等编著《西方设计：一部为生活制作艺术的历史》，长沙：湖南科学技术出版社，第 203 页）

沃尔特·克兰（Walter Crane，1845—1915）是莫里斯的主要追随者之一，他把大部分精力贡献给了工艺美术协会，传播工艺美术运动的思想，他也曾为大量书籍设计插图，后来从事建筑设计工作（图 5-9）。

威廉·理查德·莱瑟比（William Richard Lethaby，1857—1931）早年在诺曼·肖的工作室学习建筑设计，后来从事建筑历史的教学工作。他的设计范畴广泛，主要有书籍装帧设计、家具设计、建筑窗户的玻璃设计、建筑设计等。1893~1911 年，他曾创建工艺美术中心学校并担任第一任校

Walter Crane was one of Morris's main followers, and he devoted most of his energies to the Arts and Crafts association. He propagated the idea of Arts and Crafts Movement, designed illustrations for a great number of books and worked on architecture design (Illustration 5-9).

William Richard Lethaby (1857-1931) studied architectural design in Norman Shaw's studio in his early years, and later took up architecture history teaching. He had a wide range of design mainly including book design, furniture design, glass design for building windows, and architecture design and so on. From 1893 to 1911, he founded the Art and Craft Centre School and held the post of the first schoolmaster. He also

长。他还任职于英国皇家艺术学校并担任设计系教授,被授予英国皇家建筑师协会金质奖章。

三、感悟与思考

工艺美术运动源于约翰·拉斯金(John Ruskin, 1819—1900)的理论研究,以威廉·莫里斯为代表的一大批先锋设计师将拉斯金的理论研究付诸行动实现了这个理想。这场设计运动造就的艺术设计人才不胜枚举,创立了世纪行会(Century Guild)、艺术工作者行会(Art Works Guild)、家庭艺术与工业协会(Home Arts and Industries Association)、手工艺行会学校(Guild and School of Handicraft)以及工艺美术展览协会(Art and Crafts Exhibition Society),并且创办了多部专门研究纯手工艺的期刊出版物,诸如《工艺师》(Craftsman)和《陶瓷工作室》(Keramic Studio)等,在国际设计领域,这是一次卓有成效且体系完备的设计革命,具有承上启下的重要贡献,对后来的包豪斯设计产生一定影响,成为"新艺术运动"的先声,拉开了现代设计的序幕。

在创造美观与实用兼顾的设计探索中,工艺美术运动的建筑师们早年已从中国传统文化与东方建筑中吸收养分并受到启发,从文献史料可知,当时的建筑设计师曾将整座建筑采用

served as a professor at the British Royal School of Arts and was awarded the gold medal of the British Royal Technological Association.

Ⅲ. Comprehension and Thinking

The Arts & Crafts Movement was based on Theoretical research of John Ruskin (1819-1900). Represented by William Morris, a large number of pioneer designers put Ruskin's Theory into action to realize this ideal. This design movement brought up artistic design talents too numerous to be mentioned. This design movement set up the Century Guild, the Art Works Guild, the Home Arts and Industries Association, the Guild and School of Handicraft and the Art and Crafts Exhibition Society, and established multiple periodical publications specializing in pure handicraft arts such as the *Craftsman* and *the Keramic Studio*. In international design field, it was a fruitful design revolution with complete system and made important contributions of generating a connecting link between the preceding and the following. It exerted a certain impact on subsequent Bauhaus designs, became a herald of Art-nouveau Movement and initiated prologue of modern arts.

Through the design exploration of creating beauty and practical consideration, architectures of the Arts & Crafts Movement in their early years absorbed nutrition and inspired from Chinese traditional culture and oriental architecture. From the historical texts, at that time, architecture designers

木构件，吸收了东方建筑和家具设计的装饰元素特别是构件形态，讲究柱体结构的功能性和装饰性，室内家具设计亦受到中国明代家具的影响。诸如平面设计师阿瑟·海盖特·马克穆多（Arthur Heygate Mackmurdo，1851—1942）大胆借鉴东方艺术中的线条，展现以黑白线条为装饰的插图作品；奥布里·比亚兹莱（Aubrey Beardsley，1872—1898）吸收了东方艺术中的线条及装饰手法，表现唯美而纯洁的女性形象。可见，这场源于英国又盛行于欧美的设计运动，最终在东方文化、东方建筑中找到可借鉴的因素，并且东方艺术与传统文化对西方艺术家和设计家的视觉审美产生了不可胜数的影响（图5-10）。

once adopted wood constructional members for the whole building and absorbed decorative elements of oriental architecture and furniture design. In particular, in terms of component form, the functionality and decorativeness of the column structure were stressed, and the indoor furniture design was also affected by Chinese Ming Dynasty furniture. Another example was that the graphic designer Arthur Heygate Mackmurdo (1851-1942) innovatively used lines in oriental art for reference to display illustration works decorated by black and white lines. Aubrey Beardsley (1872-1898) also absorbed lines and decorative techniques in oriental art to express beautiful and unsophisticated female images. Thus it can be seen that this design movement which stemmed from Britain and prevailed in Europe and America finally sought out referential elements in oriental culture and oriental architecture. The influences of oriental art and traditional culture on the visual aesthetic appreciation of Western artists and designers were innumerable and inconceivable (Illustration 5-10).

图 5-9 由沃尔特·克兰（Walter Crane）设计的伦敦莱顿勋爵别墅的室内一景（左）
（（美）大卫·瑞兹曼著，若娴达·昂、李昶译《现代设计史》，北京：中国人民大学出版社，2012年，第120页）

图 5-10 由奥布里·比亚兹莱（Aubrey Beardsley）设计的《莎乐美》插图，运用东方艺术中的线条及装饰手法（右）
（邵宏主编，颜勇、黄虹等编著《西方设计：一部为生活制作艺术的历史》，长沙：湖南科学技术出版社，第226页）

5.3 20世纪初期西方设计思潮：新艺术运动
Trend of Western Design Thoughts from the Early 20th Century: Arts Nouveau

一、新艺术运动的文化渊源
I. Cultural Origins of Arts Nouveau

"新艺术运动"与"工艺美术运动"具有一脉相承的文化渊源。两者都对"工业化设计风格"持反对态度，都主张从自然主义、东方艺术中吸收创作的营养；两者都反对机械化的批量生产，推崇"艺术与技术"的结合。具体而言，受两次设计运动的影响，"新艺术运动"主流的设计师们往往取法自然，采用植物和动物的纹样作为主要的设计形态。当时的设计师认为大自然几乎见不到十分精确的直线条，所以，他们在设计作品中较少采用"垂直线"或"水平线"，而是以有机的曲线为中心，运用各种波浪式的、流动的、交相辉映的曲线形式，探寻一种抽象且富有象征意味的设计美学。

正如凡·德·维尔德（Henry Van De Velde, 1863—1957）所说，"工艺

The Arts & Crafts Movement and the Arts Nouveau had a direct cultural origin of succession. Both of them opposed design style of industrialization while advocate to absorb creative nutrition from naturalism and oriental arts. They also opposed mechanized mass production and held in esteem the combination of Arts and technology. Specially, affected by two design movements, the Arts Nouveau mainstream designers tended to adopt the natural law, taking vein pattern of plants and animals as the main design form. At that time, designers thought that nature could not be seen exact straight line. So, they used less vertical line or horizontal line in their design works, but took the organic curves as the center, using various wavy and flowing forms curves which enhanced each other's beauty, so as to seek a kind of abstract and symbolic design aesthetics.

As Henry Van De Velde (1863-1957) said that the Arts & Crafts Movements would open the door of modern

美术运动"开启了现代设计之门，那么，"新艺术运动"则是这股设计文化思潮的传承与发展。"新艺术运动"的成果已被大众认为达到了"青出于蓝而胜于蓝"的境地，它深入解决了产品形态与艺术文化分裂等一系列设计问题，继续向传统装饰艺术、传统文化与手工艺取经，并且继续大胆借鉴东方文化与艺术元素，更为重视人们的文化习俗与新兴设计。当"新艺术运动"传播到美国，美国文化接受了这种设计文化思想，并充分体现在建筑设计、日用器皿设计、玻璃设计等领域。法国的海报及系列平面设计最为出色，被设计界公认为现代商业广告的发源地，这股设计思潮相继在比利时、意大利、荷兰、德国、西班牙、奥地利、中欧各国乃至俄罗斯蔓延，创生出很多设计新事物。新艺术运动时期举办的巴黎国际博览会展出的各种设计产品可见，由于文化上的共性使欧洲的大部分国家与美国的设计都具有同类相求的趋向（图5-11、图5-12）。

二、新艺术运动的设计思想与设计实践

新艺术运动的设计风格各式各样，设计实践也非常广泛，一方面涉及雕塑、绘画等纯艺术领域，一方面包括建筑设计、家具设计、印刷设计、书

design. In that case, the Arts Nouveau was the inheritance and development of the design culture trend of thought. The achievement of Arts Nouveau this work was already regarded by the public as "blue is extracted from the indigo plant, but is bluer than it". It solved the product form, art culture division and a series of design problems. It continued to learn from the traditional arts, traditional culture, and handicrafts, and to boldly use oriental elements and artistic elements for reference and pay more attention to cultural custom and innovation design. When the Arts Nouveau was spread to the United States, American culture had accepted its idea of design culture, and was fully reflected in the architecture design, household utensils design, and glass design and so on. France, with most outstanding posters and series of graphic design, was recognized as the birthplace of the modern design of the commercial advertisements by design circles. This design thought had spread in Belgium, Italy, Holland, Germany, Spain, Austria, Central Europe and Russia, creating a lot of new things. During the Arts Nouveau time, all kinds of products design in The Paris Exposition Universals expo showed that most countries in Europe would have common tendency with America due to cultural commonality. (Illustration 5-11, Illustration 5-12).

II.The Design Theory and Design Practice of Arts Nouveau

The Arts Nouveau has a wide range of styles and design practices. On the one hand, it is involved in sculpture, painting and other fields of pure arts. On the other hand, it also engages in architecture design, furniture design, print

图5-11 安东尼·高蒂（Antoni Gaudi）设计的新艺术风格建筑，位于西班牙巴塞罗那
（紫图大师图典编辑部《新艺术运动大师图典》，西安：陕西师范大学出版社，2003年，第55页）

图5-12 路易斯·多米尼科·蒙塔尼（Louis Domenico Montaner）的加泰罗尼亚音乐礼堂室内设计，位于巴塞罗那
（紫图大师图典编辑部《新艺术运动大师图典》，西安：陕西师范大学出版社，2003年，第131页）

籍装帧、插图设计、染织设计、玻璃设计、陶瓷设计以及女性装饰设计等。归纳而言，追求"哥特式"和"洛可可式"的设计风格以及借鉴东方艺术与文化元素孕育了"新艺术运动"这场设计

design, books design, illustration design, textile design, glass design, ceramics design and women decoration design, etc. Sum up main thoughts of the Arts Nouveau, it pursuits Gothic design style and Rococo design style, and draws on oriental arts and cultural elements to breed three aesthetic

第5章 西方19世纪至20世纪设计文化选读
Chapter Five: Selected Readings of Western Design Culture from 19th to 20th Century

运动的三大审美取向。维克多·霍塔（1861—1947）、亨利·凡·德·威尔德（Henry Van De Veld, 1863—1957）、查尔斯·伦尼·麦金托什（1868—1928）、安东尼·高蒂（Antoni Gaudi, 1852—1926）、路易斯·多米尼科·蒙塔尼（1850—1930）是其中核心的设计师，然而，这些设计师的设计风格各有所长、不拘一格、标新立异。

维克多·霍塔（Victor Horta）是一位比利时杰出的设计师，又是"新艺术运动"的代表人物之一。他的设计生涯源于住宅设计和家具设计，他善于从象征主义艺术获得设计灵感，他提出建立一种新颖的设计语言，体现其理性主义的设计思想，他富有成就的建筑设计实例是泰西尔住宅，注重建筑每一个细部的设计，突出建筑体的肌理设计，追求完美无缺、超越历史的设计风格。但总体而论，他的诸多设计实践表明了他对传统的中世纪文化艺术与设计语言仅仅采取了间接引用和融合概括的方法，显然没有创生新的设计体系，没有突破折中主义的局面，可见，他是一位具有雄心壮志、追求完美又遵循传统的建筑师（图5-13）。

亨利·凡·德·威尔德（Henry Van De Velde, 1863—1957）最初从事绘画创作，不仅学习印象主义和象征主义艺术，还尝试过点彩派的画法。自参加"二十人社"以后，开始了他

orientations of this design movement. Victor Horta (1861-1947), Henry Van De Veld (1863-1957), Charles Rennie Mackintosh (1868-1928), Antoni Gaudi (1852-1926), Louis Domenico Montaner (1850-1930) are the core of designers. However, each of these designers has his own style, not sticking to one pattern in order to be different.

Victor Horta is an outstanding Belgian designer, and one of the representatives of the Arts Nouveau. His design career started from residential design and furniture design. He was good at getting inspiration from symbolism arts and put forward the establishment of a new design language which would embody his design thought of rationalism. His accomplished architectural design example is the residence of Tayseer. He paid attentions to the design of every detail of the architecture, highlighted the texture design of the building and pursuits perfection which beyond the historical style. In general, many of his design practices showed that he just used the method of indirect reference and fusion of the traditional culture, arts and design language in Middle Ages, and apparently did not create new design system and did not break the situation of eclecticism. We can see that he ought to be an architect with a lofty ideals and high aspirations, pursuit perfection but also follow the traditions. (Illustration 5-13).

Henry Van De Velde (1863-1957) initially engaged in painting creation, not only studied the Impression Art and Symbolism Art, but also tried the technique of Divisionism. Since joining the "20 people clubs", he started his design career. He not only adhered to the Arts Nouveau design

图 5-13 维克多·霍塔（Victor Horta）设计的画室内景，位于布鲁塞尔的索尔维饭店

（紫图大师图典编辑部《新艺术运动大师图典》，西安：陕西师范大学出版社，2003 年，第 87 页）

的设计事业。他既坚持新艺术"美术和技术结合"的设计理念，又提出"功能第一"的独到见解，他曾表示"根据理性结构原理所创造出来的完全实用的设计，才能够真正实现美的第一要素，同时也才能取得美的本质。"[1]对于各种新设计威尔德持有自己的评价标准："①产品结构设计合理；②材料运用严格适当；③工作程序明确清楚。"[2]综观而论，他倡导设计中理性的力量，而不是随意的装饰；讲究设计线条的形态与变化，使其设计更为

concept of combination of art and technology, but also put forward insight of "function first". He once indicated that "the completely practical design which is created according to the rational structure principle can truly achieve the first element of the beauty and also gain the essence of the beauty."[1] Veld had his own evaluation criteria for a variety of new designs. "Firstly, the design of product should be reasonable. Secondly, application of materials should be strict and appropriate. Thirdly, working procedure should be clear."[2] Briefly speaking, he advocated the strength of reason but not random decoration. He paid attention to the form and change of lines design which could make the design more

---

[1] 董占军、郭睿编译《外国设计艺术文献选编》，山东教育出版社，2012 年 3 月，第 152 页。（Dong Zhanjun, Guo Rui Editor.（2012）*Selection of Foreign Design Art Texts*. Shandong Education Press. p152.）

[2] 董占军、郭睿编译《外国设计艺术文献选编》，山东教育出版社，2012 年 3 月，第 154 页。（Dong Zhanjun, Guo Rui Editor.（2012）*Selection of Foreign Design Art Texts*. Shandong Education Press. p154.）

朴素简练；推崇自成一格、别具匠心且富有活力的设计作品，反对形式主义的装饰（图5-14、图5-15）。

查尔斯·伦尼·麦金托什（Charles Rennie Mackintosh，1868—1928）是一位国际公认的设计视野广阔的设计师，他的设计作品主要有海报设计、建筑设计、室内设计、灯具设计、壁挂设计，他又是一位杰出的水彩画家。他出生于格拉斯哥，日后成立了"格拉

simple and respected. He also praised highly of special, vibrant and vigorous design works and opposed formalism decoration (Illustration 5-14, Illustration 5-15).

Charles Rennie Mackintosh (1868-1928) was an internationally recognized designer with broad design vision. His design works mainly had post design, architecture design, interior design, lamp design, and wall hanging design. He was also an outstanding water color painter. He was born in Glasgow and established "Glasgow Four" design association later which would be popular throughout

图 5-14　凡·德·维尔德（Henry Van De Velde）的优秀作品《天使的注视》
（紫图大师图典编辑部《新艺术运动大师图典》，西安：陕西师范大学出版社，2003年，第205页）

图 5-15　凡·德·维尔德（Henry Van De Velde）设计的书桌
（紫图大师图典编辑部《新艺术运动大师图典》，西安：陕西师范大学出版社，2003年，第207页）

斯哥四人"设计组合,其设计声望遍及欧美国家。他的设计风格特点鲜明,采用几何图形元素,以黑白色彩为主调,运用纵横的直线布局,注重设计的均衡感。他曾倾注很多时间从事风景画的创作,大量绘画作品留存于世。他深受格拉斯哥传统建筑的影响,并开创了自己独特的建筑设计风格,他的代表性建筑设计是1907年建成的格拉斯哥美术学院大楼,这座建筑立面展现了特有的纵横格局、直线立体设计,展露出现代主义的设计风格(图5-16~图5-18)。

European and American countries. He had a distinctive design style by using geometric elements with black and white colors as the main theme, using vertical and horizontal layout and focusing on a design sense of balance. He devoted a lot of time engaged in landscape painting and a large number of paintings were retained in the world. He was deeply affected by Glasgow buildings and created his own unique architectural design style. His representative architecture design could be the Building of Glasgow Academy of Fine Arts which was built in 1907. Architectural facade of this building showed the unique vertical and horizontal pattern and liner three-dimensional design, reflecting the design style of modernism (llustration 5-16 to Illustration 5-18).

图 5-16 查尔斯·伦尼·麦金托什(Charles Rennie Mackintosh)设计的椅子

(紫图大师图典编辑部《新艺术运动大师图典》,西安:陕西师范大学出版社,2003年,第119页)

图 5-17 查尔斯·伦尼·麦金托什(Charles Rennie Mackintosh)的室内餐厅设计风格

(紫图大师图典编辑部《新艺术运动大师图典》,西安:陕西师范大学出版社,2003年,第123页)

图 5-18 查尔斯·伦尼·麦金托什(Charles Rennie Mackintosh)设计的格拉斯哥艺术学院的校舍

(紫图大师图典编辑部《新艺术运动大师图典》,西安:陕西师范大学出版社,2003年,第125页)

三、感悟与思考

1880~1910 年是新艺术运动最兴盛的时期，几乎在同一时期，象征主义也在欧洲悄然兴起，象征主义的艺术家们创作了很多表达心灵世界的作品，用视觉形象表现内心的主观感受。这种象征主义艺术深深影响了"新艺术运动"设计师们，焕发了他们的创新意识，潜移默化地体现在当时各种设计作品之中。无论从色彩、形态、线条等设计元素，还是崇尚自然主义的风格和题材的设计作品都表现出设计师们的主观感情，象征主义思潮成为设计师的灵感源泉。

我们可以想象，当时新艺术运动的影响力涵盖了建筑、家具、服装、工艺品、书籍装帧等设计领域，如影随形一般地成为文学、绘画、雕塑、戏剧、舞蹈等创作领域的主流风格，也成为欧洲以及美国的时尚文化。可见，艺术与设计犹如同胞手足，互相影响，难舍难分。

III. Comprehension and Thinking

The most prosperous period of the Arts Nouveau was from 1880 to 1910. Almost at the same time, the Symbolism had also sprung up in Europe. Symbolism arts might creat many works to express their inner world and inner feelings with visual images. This kind of Symbolism Art deeply influenced designers of Arts Nouveau. Designers of Arts Nouveau were shined creative thinking by Symbolism Art. This kind of Symbolism Art was unconsciously influenced and embodied in the various design works at that time. Not only design elements such as color, shape, and lines etc., but also design works of naturalistic style and themes all showed designers' subjective feelings. The trend of Symbolism became a source of inspiration for designers.

We could imagine that the influence of the Arts Nouveau at that time would cover all design fields such as architecture, furniture, fashion, handicraft, and book decoration, and become inseparably mainstream style of creative fields including literature, painting, sculpture, drama and dance as the shadow following the form. The Arts Nouveau also became the fashion culture in Europe and the United States. So, art and design have mutual influence like brothers and sisters and could not bear to part from each other.

5.4 20世纪西方建筑文化思潮：芝加哥学派
Trend of Western Architecture Design Thoughts from the 20th Century: Chicago School

一、芝加哥学派的由来

19世纪30年代，美国的铁路中心位于芝加哥及其临近地区，交通便利的优势使得城市人口迅速增长，人口密度增大的状况又使老村落、老社区的住房供需不足，于是在当地出现了大量简易的公共建筑与住宅建筑。这些建筑以木屋结构为主，采用"编篮式"的建造方法，密集而简陋，易遭火焚。不幸的是，1871年芝加哥遭厄火灾，芝加哥市区约8平方公里的老建筑、简易住宅被烧毁，十万火急地需要灾后重建家园。

对于多元文化背景下的城市重建，芝加哥面临着社会文化与文脉发展诸多的现实问题。芝加哥学派的建筑设计师们结合当地居民的文化习俗和生活需求，从老村落、老社区的"编

Ⅰ. The Origin of Chicago School

In the 1830s, the American railway center was located in Chicago and its adjacent areas. The urban population was growing rapidly because of the advantage of convenient transportation. The situation of density of population increase resulted in insufficient housing supply and demand among old villages and communities and then appeared a large number of simple public and residential buildings in local areas. These buildings were mainly built of like Log Cabin Structure, adopted the construction method of similar Basket Type and were intensive and humble vulnerable to fire burning. Unfortunately, the city of Chicago was hit by fire disaster in 1871, and about 8 square kilometers of old buildings and simple houses were burn down. Chicago had the most urgent need to rebuild their homes after disaster.

For urban renewal under multicultural background, the city of Chicago was confronted with many realistic problems of the development of social culture and context. Architecture designers of Chicago School combined cultural customs and life needs of local residents and they were enlightened from the similar Basket

篮式"木架构建筑群中受到启迪，重新设计出钢架铆焊梁柱，采用新型建材又在建筑风格上回归传统，给人们营造"回家"的怀旧感。

摩天大楼体现了美国建筑文化的范式。"高层建筑的特征是什么？答曰：高耸入云。这就是关于艺术本质的崇高性，具有使人激动的一面，就像风琴全部拉开的音调，非常富于感染力。接下来，它的表现必须是高和弦，是想象力的真正激发。"[①]芝加哥学派的建筑设计师们拓展了高层建筑设计，倡导"形式服从功能"的设计文化理念，从詹尼设计的曼哈顿办公楼到沙利文设计的信托银行大厦，芝加哥学派成为人们所接受的建筑文化主流。从设计文化的层面，芝加哥学派体现了一种实用文化、折中文化，再汇流成共生的西方古典建筑文化。无论是伯纳姆，还是沙利文，芝加哥学派的设计思想始终围绕着西方古典主义的审美标准，西方古典主义的文化因素一直融合在该学派的设计之中。

二、芝加哥学派的设计理论与设计实践

兴起于19世纪70年代，芝加哥学派被公认为是美国现代建筑的基

Type of old villages and communities' wooden structure complex. They redesigned beam columns of steel welding, used new building materials and returned to tradition in architectural style for people to create a nostalgic feeling of "going home".

Skyscrapers embody the paradigm of American architectural culture. What are the characteristics of high-rise building? The answer is "tower into the clouds. This is the sublimity of artistic essence. It has an exciting aspect just like the tone of the organ and touches one deeply in the heart. Next, its performance must be high chords and real spark of imagination."[①] Architects of Chicago School expanded the design of high-rise buildings and proposed the cultural concept of "Forms follow function". From the Manhattan Office Building designed by Jenny to the Trust Bank Building designed by Sullivan, the Chicago School could become the mainstream of architecture culture accepted by people. From the perspective of design culture, the Chicago School mirrors one kind of functional culture and eclecticism culture and converges into a symbiotic Western classical architectural culture. The design thoughts of Chicago School always revolves around the aesthetic standards of Western Classicism whether Burnham or Sullivan. The cultural elements of Western Classicism are always compromised among the school's design.

II. Design Theory and Design Practice of Chicago School

Rised in the 1870s, the Chicago School is acknowledged as the cornerstone of American modern architecture. The Wellborn

---

① 董占军、郭睿编译，《外国设计艺术文献选编》，山东教育出版社，2012年3月，第174页。（Dong Zhanjun, Guo Rui. (2012) *Selected Foreign Literature on Art and Design.* Shandong Education Press. P174）

石。设计师韦尔伯恩·鲁特（1850—1891）、达克·马尔·艾德勒（1844—1900）、路易斯·沙利文（1856—1924）、丹尼尔·伯纳姆（1846—1912）、威廉·霍拉伯德（1854—1923）、威廉·波宁顿（1818—1898）、马丁·罗奇（1855—1927）、威廉·勒巴隆·詹尼（1832—1907）是芝加哥学派主要代表人物。路易斯·沙利文是该学派重要的建筑设计师之一。

芝加哥学派的设计理论概括为五方面：其一，主张艺术与技术的有机结合，追求"形式服从功能"的设计原则，采用各种理性的建筑装饰设计，为西方现代主义建筑设计奠定了理论与实践基础；其二，探索"先锋式"建筑风格，其建筑风格虽源于欧洲，但力图创建一种有别于欧洲建筑，并能够真正体现美国文化、美国风貌的建筑设计体系；其三，在与西方传统建筑文化产生矛盾与冲突时，运用建筑的比例和体量展现一种力量永恒的思想，倡导将高层建筑解析为基本要素，形成富有时代精神的建筑艺术；其四，由该学派代表建筑设计师沙利文提出的"有机建筑"思想，注重整体与局部、形式与功能必须有机结合，并要考虑各种技术因素与社会文化因素；其五，提倡建筑结构方式的创新理论，往往把建筑整个中间部位作为设计要素，使建筑的

Root (1850-1891), Dark Mar Adler (1844-1900), Louis H. Sullivan (1856-1924), Daniel H. Burnham (1846-1912), William Holabird (1854-1923), William Boyington (1818-1898), Martin Roche (1855-1927) and William Le Baron Jenney (1832-1907) are main respective designers of the Chicago School. Louis H. Sullivan is one of the most important architects of the school.

The design theory of the Chicago School is summed up in five aspects as follows: The first, it advocates the combination of art and technology, pursuits the design principles of "forms follow function", uses in a variety of rational architectural design, and laid a theoretical and practical foundation for Western modernist architectural design. The second, it explores the architectural style of pioneer originated in Europe, seeks to create some design style that different from European architectures and forms one kind of architectural design system that truly reflects American culture and American architectural styles. The Third, it uses the proportions and volumes of a building to show the thought of eternal power, advocates the analysis of high-rise buildings as a basic element and form architectural art with the spirit of the times when there is contradiction and conflict with traditional Western architecture culture. The forth, the thought of Organic Architecture was put forward by the school represent architectural designer Sullivan. It not only emphasizes that the whole and the part, the form and the function must be organically integrated, but also ought to consider various technical factors and social cultural factors. The fifth, it promotes the innovation theory of architectural structure mode, tends to regard the middle part of building as a design element. The architectures' central parts,

中部与水平的建筑底部及顶层形成鲜明的对照,注重建筑的垂直形态设计。

芝加哥学派设计实践的例子举不胜举。芝加哥市区的高层建筑首次采用完善的钢框架建筑结构,发明并建造了第一台电梯运用于芝加哥的摩天大楼。很多新型建材被首次采用,整体的玻璃幕墙使新建筑大面积采光,钢材等新型建材被广泛运用于城市新兴建筑之中。"家庭保险公司大厦"是芝加哥学派的标志性设计,这是第一座用钢铁框架结构建造而成的高层建筑。这座大厦建造于1885年,正是运用了该学派设计师詹尼创立的结构原则,这座建筑的结构采用稳固而防火的金属框架,建筑外部用砖柱加以支撑,建筑内部则用铸铁的柱子加以支撑。建筑师詹尼设计的商品博览会大厦采用全石材的壁柱,上端设有古典柱头,下端建有古典基座,很好地保留了西方古典柱式的建造传统。

诸如建筑设计师路易斯·沙利文善于从西方古典建筑中汲取要素,他运用欧洲传统的砖石结构并结合建筑新材料的革新,他大胆采用古典式铸铁和石雕作为装饰形式,建成富有美国风貌的经典建筑。又如建筑设计师丹尼尔·伯纳姆以回归传统建筑风格为原则,运用独具匠心的美学观点,指导建设了蒙纳克大厦(1891年)、芝加哥哥伦比亚世界博览会(1893年),深入研究地域文化与现代设计之间的

bottom sides and top floors take shape one kind of clear contrast and pay much attention to vertical form design.

Design examples of Chicago School are too numerous to mention. High-rise buildings in Chicago city center started to use steel frame structure for the first time and the first elevator was invented to install in the Chicago Sky-Scraper. A lot of new building materials were firstly used; new buildings were lighted because of the whole glass curtain walls of new buildings; new building materials such as steel were widely used in urban new buildings. The Home Insurance Building was the landmark design of Chicago School and it was the first high-rise building constructed of steel frame structure. Built in 1885, the building was built on the structural principles of Jenny, the Chicago School designer. The structure of this building used solid and fireproof mental framework; the exterior part of this building used brick columns as support; the interior part of this building used cast iron columns as support. The Fair Building designed by the architect Jenny was made of stone wall columns, classical stigma on the top and classical base on the bottom. This building could preserve the construction tradition of the Western classical columns.

For example, architect Louis H. Sullivan built classical architectures rich in American style. He was good at drawing on elements from Western classical architecture, used traditional European masonry structures and innovation in building new materials, and boldly used classical cast iron and stone carvings as decorative forms. Another example, architect Daniel H. Burnham put forward the principle of returning to the traditional architectural style and advanced inventive aesthetic views. Burnham designed the Monadnock Block in 1891, the Columbia World Exposition in 1893, and deeply studied the relationship between regional culture and modern design. Burnham was

图 5-19 路易斯·沙利文（Louis H. Sullivan）设计的信托银行大厦
（（英）丹·克鲁克香克主编，郝红尉等译《建筑之书：西方建筑史上的 150 座经典之作》，济南：山东画报出版社，2009 年，第 211 页）

关系，他被誉为美国城市规划之父。除此以外，"芝加哥大厦""罗斯柴尔德大厦""信托银行大厦"都是出自芝加哥学派的经典建筑作品（图 5-19）。

三、感悟与思考

危机，在危险中诞生机会。芝加哥城市灾后重建的困境却给予芝加哥学派一个战胜困难、采用新材料新技术、拓展设计理念、实施设计实践、兴建现代大都会城市的机遇。芝加哥学派的建筑师们往往以一种实用而坦率的方式处理建筑设计的各种问题，从西方传统建筑文化中获取灵感，倡导现代建筑与传统文化融合创新且富有生命力的设计思想，从而在美国涌现出一大批独具本土特色的新生建筑和地标建筑。芝加哥学派展现了当时最先进的建筑设计水平。

III. Comprehension and Thinking

Crisis creates opportunities in danger. The plight of post disaster reconstruction in Chicago would give the Chicago School a chance to overcome difficulties, adopt new materials and technology, develop design concepts, put design ideas into practice and build modern bigalopolis city. Architects of Chicago School often deal with various architectural design problems in a functional and frank manner. Architects gained inspiration from Western traditional architecture culture and proposed the fusion of modern architecture and traditional culture to innovate vital design thoughts. So, a large number of unique new buildings and landmark buildings were emerged in the United States. The Chicago School was showing the most advanced architectural design level at that time.

建筑与文化有着与生俱来的亲缘关系。每一座建筑呈现出一种凝固的美，体现了文化传承的印迹，也反映了社会取向与地域文化的时代精神。芝加哥学派无论从建筑功能、建筑形式，或是建筑理念、建筑视觉各个层面推陈出新呈现了领航时代的建筑风格，为现代建筑风格的形成与发展作出了重要的贡献。值得一提的是，该学派在发展现代建筑设计理论的同时，探索西方传统文化与现代设计之间的平衡，该学派的建筑师们尊崇古典主义并且源源不断地向西方古典文化、传统古建筑学习，从中汲取创新元素以解决现代建筑设计的各种问题，既是一种现代文明的产物，又是对古典主义与传统文化的再创新。

Architecture has a natural kinship with culture. Each building presents a solidification beauty, embodies the imprint of cultural heritage and also reflects the spirit of the times of social orientation and regional culture. The Chicago School brings forth new ideas and presents an architecture style in the age of navigation no matter from building function, building types, architecture concepts or architectural visualization. It is worth mentioning that the Chicago School would like develop modern architecture theory and explore the balance relationship between Western traditional culture and modern design. Architects of the school respect classicism and continue to learn from the Western classical culture and traditional ancient architecture, which could absorb innovation elements to solve various problems of modern architectural design. The Chicago School is not only the product of modern civilization, but also re-innovation of classical and traditional culture.

思考题：
1. 请用英文概述 19 世纪西方设计文化的特点。
2. 以范文为基础，请谈谈工艺美术运动的设计理论与实践。
3. 请用英文评论"新艺术运动"的文化渊源与设计风格。

第6章　新农村改造设计
Chapter Six: Design of New Rural Reconstruction

[本章导读]

从古至今，一种房屋形制取代另一种房屋形制，一般都要历经上百年甚至上千年，才能得以实现。而今，中国的农村日新月异，仅仅三四十年的时间里，大量可见现代水泥楼房取代传统砖瓦建筑的景象。各种新型乡村建筑自然聚集而成一个个新村庄聚落形态，整体呈现出注重生活质量、追求文化生活，倡导健康理念的发展趋势。我国新农村的建设寄托着几代人的愿景，建筑设计师通过改良乡村环境以提升生活品位，美丽的乡村环境又影响着居住者的生产方式、文化生活以及现代产业的共生状态，并能够提升居住者的生活幸福感。

本章以"横街村"为实例，阐释中国新农村改造设计的新理念、新方法。主持这个项目的建筑设计师任天博士，14岁赴欧美留学，硕士毕业于美国哈佛大学建筑学专业，在国外学习和工作多年以后，他心怀乡村改造梦想而毅然回国工作，长期深扎在浙江农村，从前期设计到后期施工都亲力亲为，为村民改善居住环境而默默贡献。在他的新农村改造设计中，将工匠精神与现代设计相结合，提倡保留传统，挖掘中国传统文化的本源，把中国传统文化传承与再现出来。

6.1 任天设计师的心愿：用设计播下一颗种子
Cherished Desire of Designer Ren Tian: Sow the Seed of Hope with Design

"去年还在美国上着班，设计着城市地标，过着美国中产阶级生活的我，怎会在一年后出现在农村改造建设的工地上呢？回想起当时回国的决定，是因为一次与王澍老师的对话。我记得王老师说，'中国的文化在乡村还保留着一些有价值和有待发掘的东西，需要等人去发掘，那些老房子倒塌了还可以重建，但是文化断了就很难重拾了'。于是我发了一个愿，希望回国做一些和乡村有关的、有意义的事情。"①

"Last year, I was working in the United States designing the urban landmarks and living a life of American middle class, how did I end up at the construction site of rural reconstruction a year later? Recalled the decision of coming back home, the reason was a dialogue with the teacher Wang Shu. I remembered what the teacher Wang Shu said, 'Chinese culture in the rural villages still could retain some valuable and unearth things waiting for people to be explored. These collapsed old housed can be rebuilt, but the disconnected culture is hard to regain.' As a result, I made a wish. I hoped to come back my country and do some village related and significant things."①

缘起

Origin

用设计播下一颗种子。② 任天设计师接到这次乡村改造任务，是在 2016 年 7 月下旬。当时正值酷暑，他

Sow the seed of hope with design.② Designer Ren Tian received this rural reconstruction mission back in the late July of 2016. It was right in the intense heat of summer; when

---

① 源自任天设计师访谈录（Records come from an interview with Ren Tian designer）。
② 源自任天设计师访谈录（Records come from an interview with Ren Tian designer）。

第6章 新农村改造设计
Chapter Six: Design of New Rural Reconstruction

正好在仙居测绘一下项目基地，接到仙居县白塔镇滕书记的电话，让他去上横街村看看，帮着出出主意。那是个非常平常的小村，坐落在白塔镇政府南面的小片平原上，由于仙居县即将召开"国际绿色生态发展论坛"，上横街村将作为一个重要参观点，希望任天设计师能帮助做一些改造和提升设计。任天认为这是一个为当地百姓做点实事的好机会，就欣然答应了这个新农村改造项目的邀请。后来才意识到这个项目的时间之紧，可谓迫在眉睫、刻不容缓，仅仅一个多月的时间内需要完成所有设计，并且需要完成室内外施工，几乎是不可能的任务。但任天既然答应了，他就想必须完成好。

他就是心怀乡村改造梦想的青年设计师，中国美术学院建筑艺术学院教师任天（图6-1）。这个艰巨的项

Ren Tian was mapping the project base in Xianju County, he got the phone call from the secretary Teng at Baita Town, Xianju County and invited him to visit the Shang Heng Jie Village and provide suggestions for help. This is a very normal village located in a small flatland on the south side of Baita Town government. Because one International Green Ecological Development Forum would be held in the Xianju Town, the Shang Heng Jie Village would be considered as an important visiting site. So, the secretary Teng hoped that Ren Tian designer could do help to do some reconstruction and upgrading design. When Ren Tian thought about this good opportunity to do something useful for the local villagers, he accepted the invitation of this New Countryside reconstruction project with pleasure. He realized the time was tight only afterwards, this project was staring at me in the face and there was no time to delay, he got only one month to finish all the designs as well as all the constructions indoor and outdoor, this was a impossible mission. Yet Ren Tian had to get it done smoothly since he accepted it.

He is Ren Tian, the teacher of School of Architecture Art, China Academy of Art, the young designer with a heart of rural reconstruction dreams(Illustration 6-1). This

图6-1 14岁留学欧美，毕业于美国哈佛大学建筑学院的设计师任天，现为中国美术学院建筑艺术学院教师

目能够在如此紧张的时间内完成，取决于在过去的两年里，他曾亲自带领设计团队深扎在浙江农村，实地考察我国乡村地域文化，实地观察村民的生活习惯，为农民进行民居改造，在这里他曾遇到了一个又一个想不到的困难。

"烈日当空，任天老师在五方亭给溪石埋在瓦片里这样的设计，要给那个溪石定位，因为他觉得施工人员仅从图纸上给那个溪石定位不够准确，他就亲自搬动六七十斤重的大石头，起码有四五块，只见他亲自搬了两个多小时，准确定位之后，大汗淋漓地回到休息室。"这是一位在场的村干部所述。

因为工期时间短，还要考虑成本、造价、工期与美感的平衡，这些都是设计师面对的挑战。从前期设计到后期施工，任天设计师都亲力亲为，渐渐地与村民成了朋友。在这项改造设计完工之际，任天设计师说出了心里话："自己仿佛跟这个村已经有感情了，自己做的事情也是这个村的一部分，这个村的历史的一部分。"

任天设计师现为中国美术学院教师。他毕业于新加坡国立大学获建筑学学士学位，曾获得新加坡教育部留学基金奖学金；美国哈佛大学获建筑学硕士学位；中国美术学院博士，导

hard project was finished in such short time only because Ren Tian designer led the design team and stations deeply investigate into the village of Zhejiang in the past two years. He carried out field trip on the Chinese rural reginal culture and field observation of the living habits and customs of villagers to reconstruct the dwellings. He did experience one after another unimaginable difficulties.

"The fierce sun was hanging in the sky. The teacher Ren Tian did design work to bury brook stones into tiles in the Wu Fang Pavilion. He fixed brook stones position by himself because he felt that builders could not accurate to fix brook stones only according to drawing. He spent more than two hours on moving sixty or seventy pounds of big stones in person, at least four to five pieces of block stones until fixing on the accurate position. He was drenched in sweat and went back to the rest room." This is what a witness village cadre said.

Since time limit of the project was short and the balance of first cost, construction cost, time limit and aesthetic feelings needed to be considered, these were the challenges facing the designer. Designer Ren Tian did all he could by himself from the initial phase of designing to the later phase of construction, and gradually became a friend of the villagers. When this reconstruction design was successfully done, designer Ren Tian spoken from the bottom of his heart: "It seems that I have a connecting with this village now, everything I did is a part of this village and a part of this village's history".

Ren Tian is a teacher at China Academy of Art. He graduated from the National University Singapore and obtained his Bachelor of Architecture degree with honor. During the time, he awarded Singapore Ministry of Education Scholarship in 2000. He graduated from the Harvard

师王澍教授；苏黎世联邦高工高级访问学者。

任天荣获 2013 年 FAV Montpellier 建筑展设计建造项目竞赛第一名；2013 年南昌绿色建筑研究学院大楼设计竞赛第一名；2012 年美国哈佛住棚节设计建造竞赛第一名。他被聘为美国 MG2 建筑事务所、德国 Francis Kere 建筑事务所、瑞士 Christian Kerez 事务所、泰国 POAR 事务所、日本畏严吾事务所、西班牙 MAP 事务所、新加坡 Schaetz 事务所、新加坡 WOHA 事务所、中国中元国际建筑设计院的设计师。

任天主持设计布基纳法索首都新能源公共交通系统解决方案、尼加拉瓜布鲁菲尔德低收入学校方案设计等多项国外建筑项目、中国上横街村新农村改造设计；参与设计上海世博会中国馆、新加坡半山半岛海滨度假村、杭州淘宝总部大厦等项目。

University Graduate School of Design and obtained Master of Architecture degree. He is the PhD of China Academy of Art and his doctoral tutor is Wang Shu Professor. He was the senior visiting scholar of Eidgenoissische Technische Hochschule Zurich in Switzerland.

Ren Tian achieved the First Prize of FAV Montpellier Architecture Festival Design Build Competition in 2013. He awarded the First Prize in the Nanchang University Green Building Lab Design Competition in 2013. He awarded the First Prize in the Harvard GSD Sukkah Pavilion Competition in 2012. He was employed as architects in the MG2 Architecture in the USA, the Francis Kere Architecture in Germany, the Christian Kerez Studio Boston in Switzerland, the POAR Architect Boston in Thailand, the Kengo Kuma and Associate in Japan, the MAP Architecture in Spain, the Schaetz Architektur in Singapore, the WOHA Architects in Singapore and China IPPR Engineering Corporation.

Ren Tian presided over a number of design projects including a low cost public transportation system in Burkina Faso in Africa, a low cost public school for children in Bluefield in Nicaragua, and New Rural Reconstruction project of Shang Heng Jie Village in China. He involved in designing multiple projects in China such as the Shanghai Expo China Pavilion Competition, the BanShan BanDao Resort project in San Ya and the Taobao City Office Building in Hangzhou.

6.2 任天设计师作品实例：上横街村
Work of Ren Tian Designer: Shang Heng Jie Village

上横街村

上横街村坐落在浙江省仙居县一个大概两百户左右的小村庄，位于白塔镇南侧，毗邻神仙居旅游度假村，虽然这个村子本身没有什么自然禀赋，但是村子的"人和"是远近皆知的。上横街村的许子兵书记为上横街忙前忙后，为村民办了很多实事，是个非常有个性、说一不二的人，这也使得很多关键的项目得以实施。前几年，村中已经陆续建成了古祠堂文化礼堂，人畜分离，垃圾分类再利用等便民设施，乡村环境干净整洁，井井有条。上横街也是仙居县的明星村，已经接待了无数前来视察的宾客，这次请任天设计师担任改造设计工作，也是希望让这个村的面貌焕然一新，改造出几个亮点。

Shang Heng Jie Village

Shang Heng Jie Village is located in Xianju County, Zhejiang Province, a small village around two hundred households population, located in the south of Baita Town, and adjacent to the immortal home tourism resort. The village itself is no natural endowment but the village "people live in harmony" is well-known far and wide. Xu Zibing, the secretary of the Shang Heng Jie Village is very busy for villagers to do a lot of practical things. He is a man of great character and as good as his word. A lot of important projects were implemented because of his great efforts. A few years ago, the village were built in ancient shrine culture hall, separation of livestock and poultry, garbage recycling and other convenient facilities, rural environment clean and tidy, well-organized. Shang Heng Jie Village is also the star village of Xian Ju County. Numerous inspecting visitors were accorded a cordial reception. The invited designer Ren Tian served as the reconstruction and upgrading design work this time, also hoped to make the village take on a new look and transform a few bright spots.

第 6 章　新农村改造设计

Chapter Six: Design of New Rural Reconstruction

三角亭附近原来是一块破旧而杂乱的闲置地（图 6-2），现在变成了能为村民遮阳避雨、休闲娱乐的好地方（图 6-3、图 6-4）。

原来杂草丛生的闲置地（图 6-5），现在设计中有效地利用起来了，如今变成了能够为村民遮阳避雨、文化娱乐的五方亭，将更多的文化气息带到这里（图 6-6~ 图 6-8）。

为了能够真正改造出留得住人、富得了民的美丽乡村，任天设计师每天清早都走在乡村中进行采风，观察

Near the triangle pavilion, a tattered and ramshackle place(Illustration 6-2) has become a good place for the villagers to shelter from the rain and leisure entertainment(Illustration 6-3, Illustration 6-4).

The original weeds of the idle place(Illustration 6-5) were used effectively through design work now. Now the place could shade away from the rain and sunshine, become the Wu Fang Pavilion with cultural entertainment functions, and bring more and more things of culture here(Illustration 6-6 to Illustration 6-8).

In order to reconstruct the real "retain people, rich people" beautiful village, designer Ren Tian was walking in the village to investigate the customary, observe living habits of villagers in the

图 6-2　上横街村的三角亭改造之前的景象（实地拍摄）

图 6-3　上横街村的三角亭改造之后的景象之一（实地拍摄）

图 6-4　上横街村的三角亭改造之后的景象之二（实地拍摄）

图 6-5　上横街村的五芳亭改造之前的景象（实地拍摄）

图 6-6 上横街村的五芳亭改造之后的景象之一（实地拍摄）　　图 6-7 上横街村的五芳亭改造之后的景象之二（实地拍摄）　　图 6-8 上横街村的五芳亭改造之后的景象之三（实地拍摄）

村民的生活习惯，分不同时间段地去观察村民如何利用公共空间。看到当代村民的生活习惯，喜欢把户外作为起居空间，喜欢端着碗筷在家门口吃饭，任天设计师精心地为他们设计了具有时代感的建筑坡檐（图6-9~图6-12）。

early morning every day and observe how the villagers use public space separately in different time periods. Villagers would like use outdoors as a living space and hold bowls and chopstickers to eat at entrance of their homes. Designer Ren Tian saw the contemporary villagers' living customs and habits so he elaborately designed architectural slope eaves with a sense of the times for them(Illustration 6-9 to Illustration 6-12).

图 6-9 上横街村改造之后的建筑立面坡檐之一（实地拍摄）　　图 6-10 上横街村改造之后的建筑立面坡檐之二（实地拍摄）

图 6-11 上横街村改造之后的建筑立面坡檐之三（实地拍摄）　　图 6-12 上横街村改造之后的建筑立面坡檐之四（实地拍摄）

6.3 传统文化与设计创新
Traditional Culture and Design Innovation

上横街村改造设计的亮点正是体现在"传统文化"的运用上,通过挖掘中国传统文化的本源,以一种再创新的设计手法,将其重新展现出来。任天设计师的农居房改造与以往的有所不同,并不是大拆大建,一味仿古修旧,而是保留了古建筑原有肌理和精华部分,将工匠精神与现代风格相结合,传承中华文化,留住乡愁记忆。

上横街村的三间猪舍(图 6-13)经过设计改造,成了村民们聚会的文化展厅,定期展出一些文化传统工艺项目,诸如举办传统灯会、民间手工艺展、茶艺表演等(图 6-14、图 6-15)。

这个名叫三棵树的公共空间,原来这里只是一个破旧不堪的闲置仓库(图 6-16、图 6-17)。现在成为村民们读书、喝茶的文化空间。为了使设计改造本身不与历史断层,任天设计

Highlights of reconstruction design of the Shang Heng Jie Village are embodied the application of traditional culture. By tapping the origin of Chinese traditional culture, the reconstruction design will be shown in more innovative design techniques. Rural housing reconstruction of designer Ren Tian are different from the past because he would like retain the original texture and essence of ancient architectures and combine craftsman spirit with modern style to inherit Chinese culture and retain nostalgia memory instead of large demolition and blindly antique repair.

Three pig houses of the Shang Heng Jie Village (Illustration 6-13) were refurbished and modified to become cultural exhibition rooms for the villagers meeting, which could display some programs of traditional cultural crafts such as traditional lantern festival, folk handicraft exhibition and tea ceremony and so on(Illustration 6-14,Illustration 6-15).

Public space named Three Trees here was a dilapidated warehouse (Illustration 6-16, Illustration 6-17). Now, the Three Trees space becomes reading cultural space of reading and tea-break for villagers. In order to keep the design reconstruction from the of history fault, Ren Tian ingeniously retains many

图6-13 上横街村的三间猪舍改造之前的实景（实地拍摄）

图6-14 上横街村的三间猪舍改造之后的展厅（实地拍摄）

图6-15 上横街村的三间猪舍改造之后的文化展厅，定期展出一些文化传统工艺项目

图6-16 上横街村的碾米厂古建筑改造之前的实景（实地拍摄）

图6-17 上横街村的碾米厂古建筑改造之前的室内景象

师巧妙地保留了古建筑室内已有三百多年历史的石柱、斑驳的古墙等传统文化元素，在这些基础上增加现代设计元素，体现当代人文精神（图6-18~图6-22）。任天设计师注重"保留"古建筑的文化印记，因为在他看来其实这也是历史的一部分。

任天认为，其实这个村非常有意思，它有夯土的建筑，有石头垒的建筑，有混凝土的，还有砖的，各个时代都有，它其实是一个非常丰富的历史的叠加。我们在做的事情无非是在这个之上又叠加了一层而已。[①]诚然，

traditional cultural elements such as interior stone columns with a history of more than 300 years and a mottled ancient wall in the ancient architecture. On the basis of retaining originality work, he also applies modern design elements to embody contemporary humanistic sprit (Illustration 6-18 to Illustration 6-22). Ren Tian designer pays attention to "retaining" the cultural mark of ancient architecture because this is actually also a part of history in his view.

Ren Tian thinks, "This village is really very interesting because it would have different times historical buildings such as rammed earth buildings, stone block buildings, concrete material buildings and brick material buildings. This village is a very rich historical accumulation in reality. What we are doing is to create another layer on the top of accumulation before."[①]

---

① 源自任天设计师访谈录（Records come from an interview with Ren Tian designer）。

图 6-18 上横街村的三棵树碾米厂改造之后的文化空间之一（实地拍摄）

图 6-19 上横街村的三棵树碾米厂改造之后的文化空间之二（实地拍摄）

图 6-20 上横街村的三棵树碾米厂室内改造之后的文化空间之一

图 6-21 上横街村的三棵树碾米厂室内改造之后的文化空间之二（实地拍摄）

图 6-22 上横街村的三棵树碾米厂室内改造之后的文化空间之三（实地拍摄）

"这一层叠加"蕴涵了中国几千年的建筑文化积淀，使中国古建筑及其建筑文化得以传承与创新。

任天设计师改造乡村的情怀既出于他对乡村故土的眷恋，也出于他对中国传统文化的钟情。他认为，改造后的乡村，能够让村民们从心底里对自己的故土多一份眷恋和自信，看到这个村的年轻人会慢慢回来，让他们成为未来乡村建设的自觉参与者，让我国的年轻人对自己的传统文化更加自信。

Indeed, "This new layer" contains thousands of years of Chinese architectural culture accumulation and carries the heritage and innovation of ancient Chinese architecture culture a step forward.

Feelings of reconstruction villages come from Ren Tian' love of Chinese traditional culture deeply. Ren Tian designer is sentimentally attached to native land. He thinks, this countryside after reconstruction could let the villagers have more attachments and self-confidence for their homeland from their hearts. The young people will come back gradually and they will have their own wills to become participants of rural construction and development in the future. Young people in our country have more confidence on our own traditional culture.

6.4 设计理念与设计方法
Design Ideas and Design Methods

任天设计师的乡村改造理念

第一次做乡村改造，面对完全不在之前所学所做语境之内的建筑与构造形式，应该如何去切入这个设计成了一个难题。于是，任天想从这几个角度去介入：第一，改造过程中不给村民带来不必要的麻烦，或产生矛盾，要尽量用当地劳动力创造零工机会，让政府和村民感受到双赢；第二，只做符合当地村民的切身利益，加建与改造部分不是纯粹美学上的，而是有真正用处的，不做无用功，不要为了面目一新而搞大拆大建；第三，对老建筑的改造既要尊重历史，也要结合当代，不做修旧如旧，仿古造作的设计。唯有扎根于现实场域，问道于自然，直面对应现实给营造者带来的各种挑战，从场域与实践中探求问题的

Rural Reconstruction Ideas of Ren Tian Designer

This is my first project for rural reconstruction which would face architectures and structural forms totally different from the context of what I have learned before. How to penetrate into this design becomes a problem. Then, Ren Tian wants to intervene from these aspects: Firstly, the process of reconstruction could not cause unnecessary trouble to the villagers. In order to let the government and the villagers feel win-win, all reconstruction work ought to invite local labors as many as possible to create employment opportunities. Secondly, construction and reconstruction section are not purely aesthetic design but really useful design. Don't do any useless work. Only does design work pander to vital interests for the villagers. Thirdly, the reconstructions of ancient architecture not only respect history but also pander to contemporary trends. It is not necessary to do antique design and repair old buildings as before. Only take root in realistic field, ask the nature, and face up to all kinds of actual challenges bring to the builders to seek answers from sites and

第6章　新农村改造设计
Chapter Six: Design of New Rural Reconstruction

答案，从自然中寻找设计的灵感。①

在任天设计师的观察中发现，如今常见的乡村改造广泛缺少对乡村现状的理解与尊重，曾看到太多的乡村被改得面目全非，与历史断层，让当地人完全失去了对过去的记忆，同时，很多改造对当下的理解又不够深入，新加入的建筑元素常常是粗糙的仿古风格，这些又无法与时俱进。所以，任天所期望的乡村改造，在建筑风格上是既尊重传统又能体现当代精神。②

设计方法之一：观察与选址

带着这些理念，任天开始了设计的第一步。记得那天是早上五点，任天去现场观察，因为他想了解村民真实的生活情况，村里人的作息跟着太阳，如何使用村中空间，必须起早去看。在观察中发现，村民们通常会将室外作为他们的起居空间，端着碗筷蹲在家门口吃饭，和街对面的邻居聊天；或在村子有风凉快的地方支起小桌子，下棋或打牌，在有阴凉的地方靠着休息等。在村里走了几遍之后，任天开始慢慢确定了哪些地方是可以改造的：第一个是村口处的一个破败

practices and look for design inspiration from the nature.①

From the observation of Ren Tian designer, common rural reconstruction nowadays broadly lacks of understanding and respecting for rural existing conditions. Too many villages have been changed beyond recognition and suffered dislocation of history. The local villagers completely lose the memory of the past. At the same time, many reconstruction projects have limited understanding of rural existing conditions. New added architectural elements are always coarse antique styles without keeping up the times. So, what Ren Tian expected of the rural reconstruction and renewal is not only respect tradition but also mirror contemporary spirits in his architectural style.②

The First Design Method: Observation and Site Selection

Ren Tian began the first step in the design with these ideas. Yes, it was five o'clock in the morning. Ren Tian went to observe the village site because he wanted to understand real living situations of the villagers following the sun. He thought he must get up early to see how to use the village space. From his observation, the villagers usually like outdoor as their living space such as holding bowl and chopsticks squatting to eat in front of doorway and chatting with neighbors across the street. The villagers would like put up a small table in a cool windy place in the village to play chess or cards as their recreational activities or find in a cool place by rest and so on. After walking around this village several times, Ren Tian gradually began to determine which areas

---

① 源自任天设计师访谈录 (Records come from an interview with Ren Tian designer)。
② 源自任天设计师访谈录 (Records come from an interview with Ren Tian designer)。

的老建筑，那个建筑曾是一座老庙，里面是木结构，外部是夯土墙，以前被改为村里的碾米厂，现在已闲置多年，这里可以成为一个小茶室，供接待游客、村民或政府来客。

第二个是碾米厂边的三间旧猪舍，人们走进去还可闻到一股猪骚味，这里可以成为一个展厅，定期更新展览，植入文化设计内容。在碾米厂和猪舍对面是一排20世纪八九十年代的两层现代民房，分别是空斗墙结构和砖混结构，里面平淡也不太美观，可以做一些实用的披檐。再往里走，有一处杂草丛生的空地，有潜力改造成公共空间。村中还有一块三角形地块，现在常有老人在那里聚集，任天设想做一个帮助他们遮阳避雨的亭子。另外还有一间老房子，现在已经破败不堪，但位置又比较好，可以改造成一个村民农产品小店。

设计方法之二：设计与施工

设计的过程是艰辛而曲折的，任天设计师与工作室的三位实习生、陈蕴、危轩宇和宋雨琪几乎将设在白塔镇的工作室当成了临时的家。他们从测绘那些老房子开始，然后将其在电

could be reformed. The first was a dilapidated old building with interior timberwork and exterior loam wall in the entrance of the village, which used to be an old temple, a rice mull and had been idle for years. This old building here could be reformed a small tea room which would host guests, villagers and visitors from government.

The second was three old pig houses next to the old rice mull. People walked into the room with a pig smell before. This old pig houses here could be reformed an exhibition room, which could update exhibitions regularly and implant cultural design contents. There were a row of two story modern houses built in 1980s and 1990s opposite the sites of old rice mull and old pig houses. The row of two story modern houses was rowlock wall structure and brick-concrete structure. Inside of these houses appeared dull and no pleasing to the eye. Some functional eaves could be done. There was one vacant lot overgrown with grass and weeds if going further inside. This vacant lot here could be reformed public space. There was another triangle land block in the village often gathering elder men and women nowadays. Ren Tian envisaged designing a sheltered pavilion to help them. There was another old room in a better location, which was a ruin at that time. This old room could be reformed to a small agricultural products shop for villagers.

The Second Design Method: Design and Construction

The design process was facing a lot of difficulties and hardships. Ren Tian designer and his three trainees, Chen Yun, Wei Xuanyu and Song Yuqi almost treated the studio located in the Baita Town as a temporary home. They started to make a survey and draw of these old houses and make modeling in

脑中建模，推敲设计方案。每次方案汇报都是面对领导与村民提出各种意见，任天设计师都耐心地一一解答。

设计始终是以以人为本为出发点，充分考虑成本、造价、工期与美感的平衡，摒弃了很多建筑师追求的酷炫，任天认为，"合适""恰当"更为重要，设计上的选择也要充分考虑是否能够完整按时实施。材料上，以木结构和钢结构为主，尽量避免混凝土浇筑，因为那样时间和质量比较难以把控。整个施工过程只有一个月，任天持续参与了整个建造过程，亲自解决了无数需要现场解决的问题。比如，项目需要对多幢不同年代的建筑加盖披檐，不同年代的乡村建筑又有着截然不同的建造形式和结构体系，每一个立面上的可以做结构支撑的墙面所在位置都不尽相同，有的是空斗砖墙，有的实心砌法，有的是夯土，有的是混凝土，还有的是空心砖，所以，任天考虑到每一种特殊的情况，设计了五种不同披檐做法，来针对每一种农居建筑的结构体。

设计方法之三：设计反思

这个项目的历时虽不长，但却始终处于高密度的设计状态之中，任天设计师怀着一腔热情投入乡村建设，对上横街村的改造设计产生了一些感

computer to deliberate design schemes. Ren Tian designer patiently answered various questions and opinions to the leaders and villagers when reporting design plans each time.

Design is always people-oriented as the starting point and takes full account of the balance of prime cost, cost of construction, time limit for a project and sense of beauty. Ren Tian thinks that the "suitability" and "be to the point" are more important. The pursuit of cooling needs to be casted aside. Design choices ought to be fully considered whether the whole project could be implemented on time. Materials were mainly selected structure and steel structure instead of concrete pouring. Time and quality were difficult to be controlled if using material of concrete pouring. The whole construction process was only one month so Ren Tian continued to participate in the entire construction process and solved countless problems that needed to be solved on the spot personally. For example, buildings of different ages were needed to cap eaves according to this project. Rural buildings of different ages were completely different construction forms and structure systems. Structure supported walls on each facade were indifferent positions such as rowlock walls, solid bond walls, rammed earth walls, concrete walls and hollow brick walls. Therefore, Ren Tian took into account each particular condition and designed five types of different eaves for each type of rural residential building structures.

The Third Design Method: Design Retrospection

The duration of this project was not long. However, it always keeps in a high-density and fast-paced design status. Ren Tian designer throw himself into rural reconstruction with a warm heart and he has some feelings for the

想,设计反思是一种为未来设计项目作准备的负责态度。

设计都是遵循一种以人为本的态度,你既然要了解人,你就要首先去观察人。比如说我们在城市里面也会接触到一些商业的,或者是一些大型的项目,我们也会有这样的一个考虑。所以,我们在乡村做的很多事情,其实是一种学习,也是一种反思。①

我国乡村的现状是复杂而多元的,它虽然常常被冠以各种罗曼蒂克的想象,然而现实却通常没有想象中的那么美好。这些乡村大多只有老人与小孩,成年的儿女们都在外打工,只有过年过节才会回一趟家。平时村子里没有什么文化景观,游客可谓寥寥无几,只有杨梅节、国庆节等节假日,这里才会出现一些游人。然而,上横街村的书记和村民们非常希望把这里打造好,让这里的老百姓开起农家乐,让那些过年回乡的乡村儿女们看到村子的变化,鼓励年轻人愿意留下来继续改造家乡,真正要让自己的家乡日新月异,还是要靠村里的年轻一代。

任天设计师希望所做的一切努力是播下一颗种子,也许有一天,会在

reconstruction design of Shang Heng Jie Village. Design retrospection is one kind of responsible attitude to prepare for future design projects.

Design is always to follow people-oriented attitude. Since you want to know people, you have to observe people firstly. For example, we also come into contact with some commercial projects or some large projects in urban environment. We will also have such a consideration. So, we do a lot of works and things in countryside which actually would be not only a kind of learning but also retrospection.①

The present situation of Chinese villages is diverse. Villages are often referred to as romantic imagination but reality is usually not as beautiful as it seems. Only old people and children are living in these villages and adults are always working outside home. Grow-up sons and daughters will return home only during the Spring Festival and holidays. There was rare cultural landscape and there were usually very few tourists in this village. Here will appear some visitors only when the Waxberry Festival and National Day are coming. However, the secretaries and villagers of Shang Heng Jie Village would like reconstruct and develop their village well to open farmhouse resorts touring and let rural grow-up sons and daughters see good changes in this village when returning home for the Spring Festival. The secretaries and villagers encourage the young people who would be willing to stay in this village and reform their hometown continuously. What we really hope to change for the better day and day in own hometown relies on the younger generation in this village.

What Ren Tian designer did everything is to sow a seed. Maybe someday, the seed will sprout in the hearts of the younger

---

① 源自任天设计师访谈录(Records come from an interview with Ren Tian designer)。

年轻一代人的心中发芽，让他们成为未来建设乡村的主力军，意识到自己家乡的美丽和责任的重大（图6-23~图6-26）。

generation. The younger generation will become the main force to build up their villages in the future and they will realize the importance of their responsibility and the beauty of their hometown(Illustration 6-23 to Illustration 6-26).

图6-23　上横街村改造之后村口

图6-24　上横街村改造之后的建筑细节设计之一

图6-25　上横街村改造之后的建筑细节设计之二

图6-26　上横街村改造的原创设计师、中国美院教师任天的乡村实践基地

思考题：

1. 请用英文谈谈你对"新农村改造设计"的观后感。
2. 请思考"新农村改造设计"对你有哪些启发。
3. 建议进行一次乡村实地考察，用你的所见所闻和收集的考察资料，撰写一篇乡村改造设计实习报告。

第 7 章　建筑装置设计
Chapter Seven: Building Device Design

[本章导读]

目前，建筑装置设计已成为我国建筑学界关注的焦点。近年来，以北京奥运会、上海世博会、杭州 G20 峰会为机缘，我国兴起了以展现"中华文化"为主旨的大型公共空间装置设计的热潮。建筑装置设计可以体现一座城市的历史文脉与当代精神，它包含着功能、形态、技术等各个方面的创新。它能够反映地域文化特色、成为地标建筑，甚至具有延续历史文化、励志上进、提升城市形象等功能；它运用多元化空间转换，创生出造型新颖、形态各异的景观设计，并且能够与不同的城市环境、自然环境和人文环境相适宜；它往往伴随着新技术与传统工艺的融合与创新、新旧材料的交替与革新，继承历史文化特色，凝聚设计师的创意理念，诞生令人们耳目一新的建筑景观。

本章阐释了中国美术学院任天老师的设计实例，建筑装置设计《住棚》和《记忆山丘》是他的国外中标项目，也是获得国际赞誉的原创设计作品，作品中既蕴含了中国传统文化元素，又展现了西方文化的多元设计，通过设计理念、设计风格与设计方法的比较，使我们更系统而有条理地了解建筑装置设计的创新思维与设计运用，思考如何充分发挥它的功能与作用，使其成为发扬中华文化、美化城市环境、提升精神品质的新地标。在未来全球化设计的趋势下，进一步开阔国际视野，涌现更多中国原创设计的实例。

7.1 任天设计师作品实例：住棚设计
Work of Ren Tian Designer: SUKKAN Design

一、住棚设计的文化由来

"苏克特节"（Sukkot，也有英译为Succoth或者Sukkes）是犹太教规定的三大节日之一（图7-1），有人将它意译成中文"住棚节"。这个节日的庆祝时间是从每年秋季的希伯来历提斯利月15日（一般在公历9、10月份）开始，持续7天。在这个节日期间，犹太人住在一个用植物搭建的棚屋里，这个棚屋就叫作"苏卡"（Sukkah）（图7-2）。建造这种棚屋是为了纪念3200年以前，为了避灾而移居埃及的犹太民族被埃及人驱逐，在荒漠中流浪40年的艰苦经历，当时的犹太人就暂住在这样的临时搭建的简易棚屋里。

图7-1 住棚设计的文化由来，每年犹太人庆祝的三大节日之一

Ⅰ. Cultural Origin of SUKKAN Design

Sukkot is one of the biggest Jewish Festival(Illustration 7-1). It celebrates during the autumn of Hebrew Calendar and lasts for seven days. During the festival, Jewish people stay and eat in a small temporary shed, call Sukkah(Illustration 7-2). The Sukkah is to commemorate the event of Exodus when Jewish people were evicted from Egypt 3200 years ago and lived in the desert for 40 years. They lived in this kind of temporary structure.

二、"苏卡"设计项目:《漫长的旅程》

"苏卡"(Sukkah)是纪念犹太人脱离埃及的奴隶之后,在沙漠漂泊40年,移居简陋住处经历的一种回忆。对于这个"苏卡"项目,展馆装置设计的寓意在于回顾这段历史的漫长旅程,也如同潮汐涨落一般让人们再回首那段漫长的犹太传统的住棚构筑设计历程。开启历程这次旅行,先进入一条狭长的走廊,随后逐渐地提升,这正是人们进入下一个主要的大空间而预热筹备的地方。在走廊尽头,这个"预热"空间转而延伸到漫长而宽阔的休息空间,人们可以在此尽情欢乐与休闲。这个"转角"结构设计在限定边界的空间里营造了一条最漫长的路径。漫步旅行本身也是一种休闲放松的过程。从剖面图来看,"苏卡"(Sukkah)住棚构筑采用了分段框架结构的标准木材外观。这种构造的柱型比例创生了一个被长久持续笼罩着"身临其境"的围屏,这恰是建筑结构本身的效果体验并且考虑到人们可以从不同程度与户外进行视觉交流(图7-3)。

II. A Sukkah Project: Long

The Sukkah is a reminiscence of fragile dwellings in which the Israelites dwelt during their 40 years of travelling in the desert after the Exodus from slavery in Egypt. For this Sukkah project, the pavilion is designed as a long journey which recalls this history. It also tides back to this long Jewish tradition of Sukkah construction. The journey starts with entering to a narrow corridor that gradually ascends and this is where one prepares one-self to enter the main space. At the end of the corridor, the space turns and descends into the long resting space, where people rest and rejoice. This turning configuration creates the longest path within the given boundary. The walking journey itself is also a process of resting. The Sukkah is constructed with standard wooden profiles of sectional framed structure. This tectonic modulation creates a long continuous wrapping screen which is the structure itself and allows for different degree of visual connection with the exterior(Illustration 7-3).

图7-3 住棚设计外部的实地景观
(原创设计师:中国美术学院教师任天)

图7-2 住棚节的传统文化,其间人们在棚内用餐、休息、聚会,欢庆节日

7.2 创意构思与设计方法：住棚设计
（图 7-4~ 图 7-15）

Creative Ideas and Design Methods: SUKKAN Design
(Illustration 7-4 to Illustration 7-15)

图 7-4　住棚设计的原理图
（原创设计师：中国美术学院教师任天）

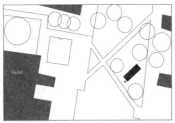

图 7-6　住棚设计的结构模型图之一（XYZ 坐标）
（原创设计师：中国美术学院教师任天）

图 7-7　住棚设计的结构模型图之二（XYZ 坐标）
（原创设计师：中国美术学院教师任天）

图 7-5　住棚设计的总平面图
（原创设计师：中国美术学院教师任天）

图 7-8　住棚设计的结构模型图之三
（原创设计师：中国美术学院教师任天）

第7章 建筑装置设计
Chapter Seven: Building Device Design

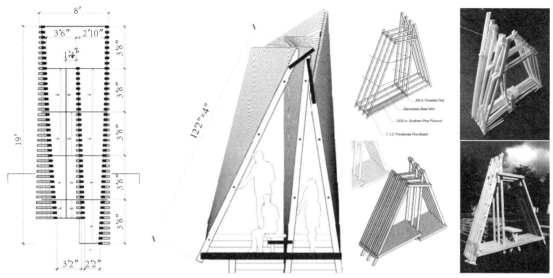

图7-9 住棚设计的平面图　图7-10 住棚设计的剖面图　图7-11 住棚设计的结构展示图
（原创设计师：中国美术学　（原创设计师：中国美术学院教师任天）　（原创设计师：中国美术学院教师任天）
院教师任天）

图7-12 住棚设计的序列细节展示图
（原创设计师：中国美术学院教师任天）

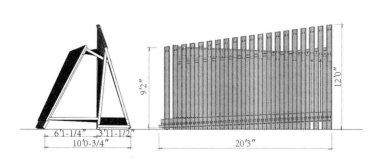

图7-13 住棚设计的正面和侧面的立面图
（原创设计师：中国美术学院教师任天）

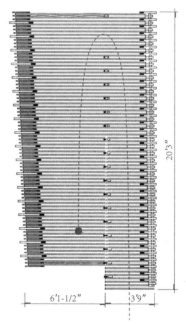

图7-14 住棚设计平面交通路径图
（原创设计师：中国美术学院教师任天）

每一段排架间距由三个重复的部分组成。结构原理是一致的，但是设计成不同的长度。前方与后方的主要空间用半透明的织物包围起来。其顶部设有枝状物。三个重复部分的每一项由五个水平的构件交织在一起，等间距的交替变换。为了减轻施工程序，这个住棚展馆被分为六个半的排架间距，分段建造（图 7-16）。

因为不同的间距，安置空间以防止各个构件向一侧移动。这种空间安置也给予建筑更好的横向稳定性的结构（图 7-17、图 7-18）。

Each bay is made of three repeated parts. The logic of construction is the same, but with different length. The front and back end of the main space is enclosed with translucent fabric. The topped with branches. Each of the three parts is weaved by five planes of members, at alternating intervals. To ease the construction process, the pavillion is broke down into six and half bays to build separately(Illustration 7-16).

Due to the different spacing, spaces are placed to prevent members to move sidewards. They also provide lateral stability to the structure(Illustration 7-17, Illustration 7-18).

图 7-15　住棚设计建造过程图（实地拍摄）
（原创设计师：中国美术学院教师任天）

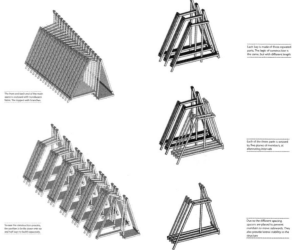

图 7-16　住棚设计的创意构思步骤展示
（原创设计师：中国美术学院教师任天）

第 7 章 建筑装置设计
Chapter Seven: Building Device Design

图 7-17 住棚设计内部的实地景观
（原创设计师：中国美术学院教师任天）

图 7-18 住棚建成的实地景观
（原创设计师：中国美术学院教师任天）

7.3 任天设计师作品实例：记忆山丘
Work of Ren Tian Designer: Memory Scape

文化元素与文化内涵

从传统习俗来看，《记忆山丘》是一座映射出传统中国文化底蕴的园林景观的自然模型缩影。一片自然之地被重新设计成一座微缩型尺寸的标记象征更广阔的自然，以标牌作为特性符号，产生对想象力或是真实自然的回忆。

中国文化元素：设计的形式来源于东方的枯山水，中国文人善在见方之地掇山理水，以寄情自然，同样的概念也发生在法国，作品中汲取了法国普罗旺斯薰衣草田的形式，将东西文化融入其中。同时，喜怒哀乐是人类共同的情感，跨越年龄，跨越时代，人们在互动的过程中释放了各自心中最动情的瞬间，也让这个瞬间在公共建筑的装置中凝固，最终成为跨越时空、跨越文

Cultural Elements and Cultural Connotation

Formally, the mamoriscape echoes the dried miniature landscape gardens in the traditional Chinese culture, where a piece of nature is recomposed in a miniature scale to signify the longing for the greater nature, function as a sign, evokes the memory or imagination or the real nature.

Elements from Chinese Culture: The design form comes from the Chinese Scholar Garden Rock. Chinese literatis are experts in piling up rocks and managing water in the square places and abandon themselves to nature. Exactly, the same concept also takes place in France. This work draws the form of lavender fields in Provence, France and integrates Eastern and Western cultures in harmony. At the same time, pleasure, anger, sorrow, joy – the passions are common emotions of mankind across the ages and the times. People release the most emotional moments in their hearts in the process of interaction. The most emotional moments have

化、属于所有参与者的景观（图 7-19、图 7-20）。

also curdled in the public building device in this work. It is eventually the landscape beyond times, space and cultures that belongs to all participants(Illustration 7-19, Illustration 7-20).

图 7-19 《记忆山丘》的中国传统文化意象之一
（原创设计师：中国美术学院教师任天）

图 7-20 《记忆山丘》的西方传统文化意象之一
（原创设计师：中国美术学院教师任天）

7.4 设计理念与设计方法：记忆山丘
Design Concepts and Design Methods: Memory Scape

在设计方案中，这是唯一一种柏拉图式形态的圆形循环，既不带有任何方向性的暗示，也不会与任何棘手的边缘发生冲突。它也巧妙地处理了方形庭院的四周，不存在任何偏倚（图7-21）。

《记忆山丘》的设计具有三维坐标的指示符号，即年代X，年龄Y，情绪Z。参观者可以将他们的记忆或者愿望定位安插在这些坐标轴上（图7-22~图7-26）。

In plan, it is a circle shape as circle is the only platonic shape that does not infer any directionality or encounter the awkwardness of edge. It also address the four sides of the square courtyard equally without placing any bias(Illustration 7-21).

The mamoriscape is indexed with three axises, year (X), age (Y), and mood (Z). Visitors may want to place their memory or wish according to those axises(Illustration 7-22 to Illustration 7-26).

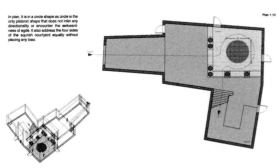

图7-21 《记忆山丘》的总平面设计图之一
（原创设计师：中国美术学院教师任天）

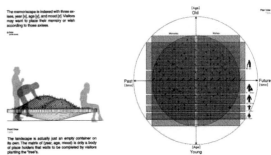

图7-22 《记忆山丘》的总平面设计图之二
（原创设计师：中国美术学院教师任天）

第7章 建筑装置设计
Chapter Seven: Building Device Design

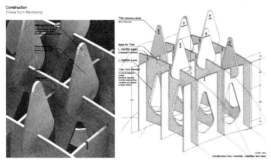

图 7-23 《记忆山丘》的构造设计图
（原创设计师：中国美术学院教师任天）

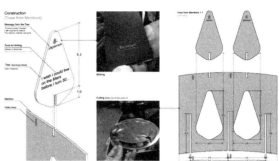

图 7-24 《记忆山丘》的构造细部图解
（原创设计师：中国美术学院教师任天）

图 7-25 《记忆山丘》的局部构造模型
（原创设计师：中国美术学院教师任天）

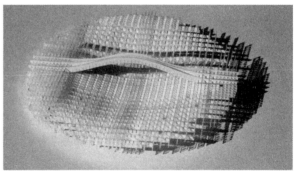

图 7-26 《记忆山丘》的整体构造模型
（原创设计师：中国美术学院教师任天）

实际上，这种景观仅仅是一片专属于它的空白的容纳空间。这种年代、年龄和情绪的形成层面仅仅是支托物指定位置的主体部分，期待着参观者亲自栽种以塑造完善这一棵棵树（图7-27~图7-33）。

The landscape is actually just an empty container on its own. The matrix of (year, age, mood) is only a body of place holders that waits to be completed by visitors planting the "trees"(Illustration 7-27 to Illustration 7-33).

图 7-27 《记忆山丘》的建造现场（实地拍摄）
（原创设计师：中国美术学院教师任天）

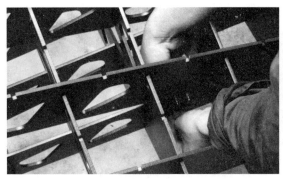

图7-28 《记忆山丘》的建造过程
(原创设计师：中国美术学院教师任天)

图7-29 《记忆山丘》的现场实景
(原创设计师：中国美术学院教师任天)

图7-30 《记忆山丘》实景与现场观摩
(原创设计师：中国美术学院教师任天)

图7-31 《记忆山丘》实地观摩深受观众喜爱
(原创设计师：中国美术学院教师任天)

图7-32 《记忆山丘》作品细部特写
(原创设计师：中国美术学院教师任天)

图7-33 《记忆山丘》原创设计作品深受欢迎
(原创设计师：中国美术学院教师任天)

思考题：

1. 请思考作品《住棚》对你有哪些启发。
2. 结合作品《记忆山丘》，谈谈建筑装置设计的文化内涵。
3. 建议结合中国文化主题，构思一项空间装置设计作品并制作模型。

第 8 章　工业产品设计
Chapter Eight: Industry Production Design

[本章导读]

从现存文献考辨可知，早在两千多年前（周）卜商撰写的《子夏易传》一书中最早提出"造物"一词。其后，在《列子》《庄子》《释名》等多部古典文献中分别阐述了我国古代"造物"思想。伴随着"造物观念"的产生及运用，我国古代能工巧匠们创造了各种工具、制造了各式器物。然而，"造物"思想正是当代"设计"的原点。顺应国际化发展的大趋势，同行同类设计产品之间的材料、工艺、质量、功能、科技水平的差异已经日益缩小，消费者很难在短时间里辨别出产品高低而作出选择，唯有产品设计内涵的"文化特色"，诸如设计是否符合人们的生活习俗，是否体现人们的文化品位，这些成为产品设计成功的重要因素。

本章展现了浙江工业大学设计艺术学院朱昱宁老师的《八宝吉祥扇》获奖设计作品、毕业于浙江工业大学和美国帕森斯设计学院工业设计专业郑昱设计师的《延绵·书桌设计》和《蒸汽加热餐具组合》的产品设计实例，两位设计师长期致力于工业产品设计的研发，将传统文化元素与现代科技相结合，阐释当代急需的将工业产品设计与传统文化元素融合创新的新构思、新产品。通过这些设计实例有助于我们从传统文化中寻找设计灵感、开阔设计思路、思考设计定位、形成原创的设计风格，并且结合现代工业先进技术不断开发新设计。

8.1 朱昱宁设计师的感言：设计传承文化
Understanding of Designer Zhu Yuning: Design Inherits Culture

"设计传承文化"[①]朱昱宁设计师现任教于浙江工业大学设计艺术学院工业设计系，他本科和硕士都毕业于浙江工业大学设计艺术学院工业设计专业。他担任浙江省工业设计技术创新服务平台设计部部长、杭州飞神工业设计有限公司设计总监。他长期从事一线工业设计开发，传统手工艺行业的工业产品设计介入与产品推广策划（图8-1）。

Design inherits culture.[①] Designer Zhu Yining graduated from the School of Design, Zhejiang University of Technology and got his Bachelor Degree and Master Degrees of Industry Design. He is the teacher of the Department of Industry Design of School of Design, Zhejiang University of Technology. He is serving as the Industry Design Minister of Zhejiang Province Industry Design Technology Innovation Service Platform. He is the design director of Hangzhou Fei Shen Industry Design Co., Ltd. He has long been engaged in the development of front-line industrial designs, especially the industry product design, design intervention, and product promotion planning of traditional handicraft industry (Illustration 8-1).

图8-1 工业产品设计师：朱昱宁（现为浙江工业大学设计艺术学院教师，浙江省工业设计技术创新服务平台设计部部长，杭州飞神工业设计有限公司设计总监）

---

① 设计感言源自设计师朱昱宁访谈录（Records come from an interview with Zhu Yuning designer）。

8.2 朱昱宁设计师作品实例：八宝吉祥扇
Work of Zhu Yuning Designer: Eight Auspicious Symbols Fan

《八宝吉祥扇》设计作品曾获得"浙江省工业设计大赛优秀奖"。获奖设计师：朱昱宁（浙江工业大学）、孙亚青（杭州王星记扇厂）、王冰（图8-2）。

Design work titled the "Eight Auspicious Symbols Fan" was gained of The Excellent Award of Zhejiang Province Industry Design Competition. Award-winning Designer: Zhu Yuning from Zhejiang University of Technology, Sun Yaqing from the Hangzhou Wangxingji Fan Co., Ltd. and Wang Bing (Illustration 8-2).

图8-2 《八宝吉祥扇》获得"浙江省工业设计大赛优秀奖"
（原创设计师：朱昱宁（浙江工业大学设计艺术学院）、孙亚青（杭州王星记扇厂）、王冰）

一、《八宝吉祥扇》的文化内涵

清光绪元年（1875年），江南制扇名匠王星斋开创了"王星记扇庄"，正是杭州王星记扇厂的前身。中国的扇文化具有深厚的文化底蕴，制扇技艺被列入国家非物质文化遗产保护名录。杭州王星记扇厂是中国唯一驰名中外的传统品牌扇子，被誉为"贡扇"，也是国家商务部认定的"中华老字号"。

王星记的传统工艺扇构造精制、技艺精湛，凝聚中国艺术之精华。经典的中国传统黑纸扇曾在巴拿马博览会、西湖万国博览会上屡次获奖，拥有"东方瑰宝"之美称。杭州佛教起源于东晋，五代时期流传较广，南宋时期达到顶峰，与杭州的地域文化相得益彰。佛教八宝，也叫八吉祥，是象征佛教的八种图案纹理，作为图案装饰历史悠久，八宝寓意吉祥，作为主要纹饰被广泛运用在中国传统工艺品之中（图8-3）。

I. Cultural Connotation of the Eight Auspicious Symbols Fan

The fan master of regions south named Wang Xingzhai established the "Wangxingji Fan Shop" in 1875 in the Qing dynasty, which was the predecessor of Hangzhou Wangxingji Fan Industry Co., Ltd. Chinese fan culture has profound cultural inside information. The fan craft was recorded into the national intangible cultural heritage protection list. In history, the Wangxingji Fan was the tribute fans. Now, the Hangzhou Wangxingji Fan Industry Co., Ltd. is the China's time-honored brand approved by the Ministry of Commerce of the People's Republic of China.

Traditional Wangxingji's fans were won high reputation in the circle for the features in material, technological process and decoration and condense the essence of Chinese art. The classical Chinese traditional Black Paper Fan has repeatedly won prizes at the Panama Expo and the West Lake World Fair, which have the reputation of "oriental treasure". Buddhism in Hangzhou began in the Eastern Jin dynasty, bloomed in the Five dynasties and flourished in the South Song dynasty. The Buddhism and Hangzhou reginal culture made the old and new contrast and complement each other. The eight treasures of Buddhism, or named eight auspicious symbols could become eight pattern textures of Buddhist symbols and have a long history as pattern decorations. The eight auspicious symbols contained good meaning of luck and happiness so were used widely in Chinese traditional handicrafts as main pattern decoration (Illustration 8-3).

八宝吉祥扇
单根扇骨长23cm,
宽2.2cm, 28根

图8-3 《八宝吉祥扇》设计作品展示图
（原创设计师：朱昱宁（浙江工业大学设计艺术学院）、孙亚青（杭州王星记扇厂）、王冰）

二、《八宝吉祥扇》的设计特色

设计师将"八宝"纹理与杭州"王星记"手工精制的乌木扇相结合,推出了"八宝吉祥扇"这一全新的佛教礼品扇。这件作品曾作为杭州市的城市文化礼品馈赠海内外嘉宾。

《八宝吉祥扇》的设计造型优美、精制典雅,以佛教礼品扇赠送友人,可以表达美好祝福;不仅适合于随身携带,又适合置于室内陈设,寓意趋利避害、吉祥如意。本系列乌木扇设计是通过计算机辅助工业设计开发的扇子高效设计系统开发而成,相比传统手工艺的繁复耗时,形成了标准化、现代化的扇子设计研发体系(图8-4)。

II. Design Features of the Eight Auspicious Symbols Fan

Designer combined the two aspects, one was the vein and pattern of "Eight Auspicious Symbols", the other was the hand crafted ebony fan of Hangzhou Wangxingji Fan Co., Ltd. He designed this brand new Buddhist gift fan titled Eight Auspicious Symbols Fan. This work was honored as the City Cultural Gift of Hangzhou to guests both at home and abroad.

The work of Eight Auspicious Symbols Fan was refined and designed beautifully. Buddhist gift fans as the gifts could express good wishes to friends. The fans were not only suitable for carrying but also suitable for interior furnishings. The implied meaning was to pursue good fortune and avoid disaster as lucky as desired. This series of ebony fan design were designed by computer aided industry design, the efficient and high-speed design system. Compared to the traditional handicraft disadvantages of complicated and time-consuming, the standard and modernized fan design and R & D system have been formed (Illustration 8-4) .

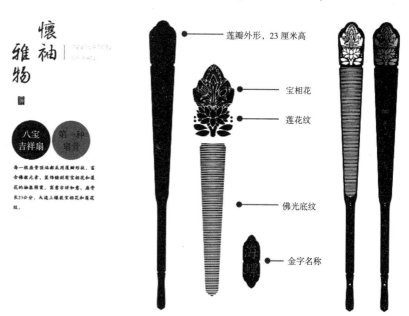

图8-4 《八宝吉祥扇》作品设计图
[原创设计师:朱昱宁(浙江工业大学设计艺术学院)、孙亚青(杭州王星记扇厂)、王冰]

8.3 郑昱设计师的心得：设计源于生活
Feeling of Designer Zheng Yu: Design Comes from Life

设计心得："设计源于生活。"[①]郑昱毕业于浙江工业大学设计艺术学院工业设计系，得到卢纯福、朱上上、黄薇、张露芳、刘肖健、傅晓云、朱意灏、徐冰教授的精心指导。她毕业后赴美国深造，研究生毕业于美国帕森斯设计学院工业设计专业。她在国内外多家科技研发企业、产品设计公司担任专业设计师，现在已成为一位具有国际化设计视野、思维敏捷并且能够独立承担各种产品设计项目的优秀青年设计师（图8-5）。

Design Feeling: "Design comes from life."[①] Zheng Yu graduated from School of Design, Zhejiang University of Technology and got her Bachelor Degree of Industry Design. She got meticulous guidance and instruction from Professor Lu Chunfu, Zhu Shangshang, Huang Wei, Zhang Lufang, Liu Xiaojian, Fu Xiaoyun, Zhu Yihao and Xu Bin during undergraduate study time. She continued her study in America after graduation. She graduated from the Parsons School of Design in USA and got her Master Degree of Industry Design. She is serving as a professional designer in technology research and development enterprises and products design companies both at home and abroad. She has now become an excellent young designer with international design vision, who could independently undertake all kinds of product design projects and be quick intellect (Illustration 8-5).

图8-5　毕业于浙江工业大学设计艺术学院、美国帕森斯设计学院工业设计专业的设计师郑昱

[①]　摘自郑昱设计师访谈录（Records come from an interview with Zheng Yu designer）。

8.4 郑昱设计师作品实例：延绵·书桌设计
Work of Zheng Yu Designer: Stretch Long & Continue Forever Desk Design

《延绵·书桌设计》是一款融合中国传统文化的现代创意书桌。桌面的小型景观设计集功能性、美观性、舒适性于一体，从而有效地缓解了繁忙紧张的工作学习带来的压力（图8-6、图8-7）。

Stretch Long & Continue Forever Desk Design is one kind of contemporary creative desk combining with traditional Chinese culture. The small landscape design of desktop integrates functionality, beauty and comfort. Thus, the stress coming from busy and tense work and study could be effectively relieved (Illustration 8-6, Illustration 8-7).

图8-6 融合中国古典园林元素的作品《延绵·书桌设计》 图8-7 《延绵·书桌设计》的细节展示图
（原创设计师：郑昱） （原创设计师：郑昱）

一、设计文化内涵

《延绵·书桌设计》的桌面小型景观设计将海绵材质的山石造型与木质书桌结合在一起,将中国古典园林艺术"叠山理水"的营造手法运用到书桌设计之中,中国传统园林建筑讲究统一中求变化,变化中不失统一的布局,善于运用复杂的曲线、不规则的外形造景,经过实验采用海绵材质恰到好处地塑造了富有中国诗画意境的微型园林景观。这款设计还巧妙地增设了一些绿色植物盆景,营造了一种仿佛置身于大自然的意境空间。

二、设计定位

该项设计的用户定位为工作节奏快的脑力劳动者。这款书桌设计的舒适性能好,通过改善现代脑力劳动者的工作环境,有益于提高工作效率,有益于坚守长时间工作,有益于接受新事物、新挑战。

该项设计的环境定位适合于创意工作空间和家庭书房。一般情况下,对工作空间环境的要求分为严肃型、创意型。对于从事创意型工作的人士,这款书桌设计最为适宜,它不仅带给人们亲近自然、轻松自如的空间享受,并且能够激发人们的想象力和创造力。在家庭书房内采用这款书桌能够创设一种亲切柔和的氛围。

Ⅰ. Design Cultural Connotation

The desktop small landscape design of Stretch Long & Continue Forever Desk Design integrates rockery modeling of sponge materials with wooden desk together. It uses the construction techniques of the "Pile up rocks and manage water" Chinese classical garden art into desk design. Chinese Landscape architecture emphasizes on the overall arrangement of changes in unity and unity in change, well utilizes complicated curves and irregular appearances to embody landscape. Through experiments, it perfectly uses sponge materials to shape small garden landscape with rich artistic conceptions of Chinese poetry and painting. This design is also ingeniously add some green plants potting to create an artistic conception space as if being in nature.

Ⅱ. Position Design

User Position of this design work is suitable for mental workers with fast pace life. Comfort performance of this desk design is very good. The desk design is beneficial to improve the work efficiency, stick longer work time, and accept new things and challenges through improving the work environment of modern brainworkers.

Environmental position of the design is suitable for creative workspace and home study room. In general, the requirements of work space could be divided into serious and creative types. This desk design is the most suitable for people engaged in creative work. It not only brings people close to nature and enjoy relaxing and free space, but also could inspire people's imagination and creativity. This kind of desk design could create a sense of tenderness when adopting in home study room.

三、设计方法与材料实验

Ⅲ. Design Methods and Material Experiment

该书桌的材质分为海绵和木质两个部分。海绵材质柔软且富有弹性，上色和塑造成为具有坚硬外观的山石效果，产生一种强烈的对比，增强视觉冲击力。海绵材质主要用

Material of the desk is divided into sponge and wooden two parts. Sponge material is soft and elastic, produces a strong contrast and enhance visual impact when colored and shaped into effect of one kind of solid rock appearance. Sponge material is mainly used to take in. Single sponge material area is slightly

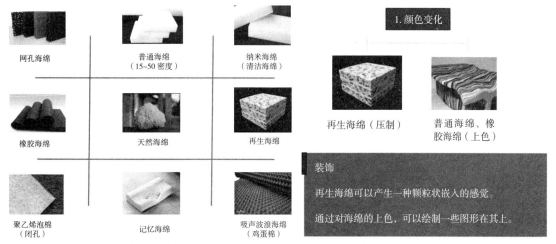

图 8-8 《延绵·书桌设计》的材料种类与选材过程
（材料实验者：郑昱）

图 8-9 《延绵·书桌设计》关于颜色变化的材料实验
（材料实验者：郑昱）

天然海绵（表面形状变化）

橡胶海绵（易切割，不易变形）

吸声波浪海绵（造型独特）

普通海绵、网孔海绵（内部）

聚乙烯泡棉

装饰
可以直接使用海绵进行切割后制作家具。
可以将海绵附着在一些起支撑作用的材质上做出一些造型。

灯具制作
由于海绵多孔，可产生一种朦胧美。
同时可以利用其不规则的外表来突出灯具的独特性。
可以将海绵附着在透明支架上来制作灯罩造型。

图 8-10 《延绵·书桌设计》关于造型变化的材料实验
（材料实验者：郑昱）

图 8-11 《延绵·书桌设计》关于透光性的材料实验
（材料实验者：郑昱）

于收纳，单边（根据使用者的书写习惯，左边或右边）海绵材质的面积略微大于桌面，这样设计充分考虑到人们书写时的需求，起到很好的支撑手肘的作用，利于书写（见图 8-8~ 图 8-15 ）。

larger than other desktop area (according to user's writing habits, left or right) because this kind of design could fully consider people writing requirements, play a good supporting role in the elbow and conducive to writing (Illustration 8-8 to Illustration 8-15).

图 8-12 《延绵·书桌设计》关于触感的材料实验
（材料实验者：郑昱）

图 8-13 《延绵·书桌设计》关于表面纹理的材料实验
（材料实验者：郑昱）

图 8-14 《延绵·书桌设计》关于材料组合的实验
（材料实验者：郑昱）

图 8-15 《延绵·书桌设计》关于海绵上色的材料实验
（材料实验者：郑昱）

四、设计方法与设计流程
（图 8-16~图 8-18）

IV. Design Methods and Design Cycle (Illustration 8-16 to Illustration 8-18)

图 8-17 《延绵·书桌设计》体现设计方法的产品草图之一
（原创设计师：郑昱）

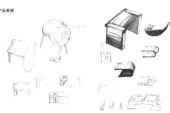

图 8-16 《延绵·书桌设计》设计思维与分析图
（原创设计师：郑昱）

图 8-18 《延绵·书桌设计》体现设计方法的产品草图之二
（原创设计师：郑昱）

五、《延绵·书桌设计》的方法与流程（图 8-19、图 8-20）

V. Method and Cycle of the Stretch Long & Continue Forever Desk Design (Illustration 8-19, Illustration 8-20)

材料调研 → 市场调研 → 用户调研 → 材料实验 → 设计定位 → 设计方案 → 模型制作 → 产品展示。

Material Investigation → Market Research → User Survey → Material Experiment → Position Design → Design Proposal → Model Building → Work Presentation.

图 8-19 《延绵·书桌设计》体现设计方法的制作流程图
（原创设计师：郑昱）

图 8-20 《延绵·书桌设计》设计展示图
（原创设计师：郑昱）

8.5 郑昱设计师作品实例：蒸汽加热餐具组合
Work of Zheng Yu Designer: Half & Half

《蒸汽加热餐具组合》是一款融合中国传统饮食文化的餐具组合设计。从蒸锅部分到煮锅部分、盛水容器部分、基座部分等每一个部件都凝聚了设计师对生活的感悟。这组设计以其便捷合理、精美实用的特点赢得了都市青年朋友的青睐（图8-21、图8-22）。

The Half & Half is the combination design of tableware of Chinese traditional food culture. From the steamer part, the boiler part and the water container part to the base part, each component condenses the designer's understanding of life. This set of design exhibits convenient, reasonable, exquisite and practical characteristics and wins favour in urban young friends' eyes (Illustration 8-21, Illustration 8-22).

图8-21 《蒸汽加热餐具组合》设计展示图
（原创设计师：郑昱）

图8-22 融合传统中国饮食文化的作品《蒸汽加热餐具组合》设计展示图
（原创设计师：郑昱）

一、文化本源与设计灵感

《蒸汽加热餐具组合》的创意来源于中国传统饮食文化。在中国饮食文化历史上,周·卜商撰写的《子夏易传》写道:"君子以饮食宴乐,……"① 可见,中国古代文人雅士注重饮食"五味调和",使菜肴的色彩、口感、营养各有千秋。同时,中国饮食文化也成为君子交友雅集、抒发胸臆的巧妙方式(图 8-23)。

中国传统菜系的色香味俱全,然而,传统菜肴的烹饪方式各不相同、品类丰富,本组设计的灵感源自中国传统饮食的"蒸"和"煮",以健康美味的烹饪方式为着眼点,汲取起源于我国汉代"蒸笼"的特点、功能与造型,将传统饮食文化与当代产品设计结合创新(图 8-24、图 8-25)。

Ⅰ. Origin of Culture and Design Inspiration

The originality of the Half & Half comes from Chinese traditional food culture. In the history of Chinese food culture, Bo Shang wrote the book titled *Zi Xia Yi Zhuan* in the Zhou dynasty and said that a man of noble character would entertain the distinguished guests to prepare nice food and drink. So, Chinese ancient literatis stress the Diet of Five Flavors Harmony and make the dishes come in different colors, tastes, and nutrition. At the same time, Chinese food culture also becomes ingenious ways for gathering friends, expressing minds and communication (Illustration 8-23).

The color, the aroma and the taste of Chinese traditional cuisine is available all varieties. However, cuisine methods of traditional dishes are diverse from each other and have rich categories. As the starting point of healthy, delicious cooking way, this design inspiration comes from Chinese traditional diet "steaming" and "cooking", draws characteristics, function and modelling of the "cooking steamer" from the Han dynasty and combines the traditional diet culture with the contemporary product design (Illustration 8-24, Illustration 8-25).

图 8-23 中国传统饮食色香味俱全,品类齐全、五味调和、营养丰富

① 周·卜商撰《子夏易传》,见清代通志堂刻本。

图 8-24 制作中国传统菜肴"蒸""煮"的烹饪方法伴随着中国传统饮食习惯

图 8-25 "蒸笼"起源于汉代,传承至今已有两千多年历史。本组设计从中国传统饮食器具中汲取设计灵感

二、设计方法与探索实验

Ⅱ. Design Methods and Exploration Experiment

本组设计的探索实验一方面在于研究产品造型,如何有效地将不同材质的烹饪器具"叠加"与"组合";一方面需要反复进行材料实验,运用适合于蒸煮的材料制作出犹如"蒸笼"一般的中式烹饪器具,将中国传统饮食习惯与先进的现代化电子锅具设计相融合(图 8-26、图 8-27)。

On the one hand, the Exploratory Experiment of this design is about product modeling and how to effectively pile up and make up cooking utensils of different materials. On the other hand, this product requires doing material experiments repeatedly, using suitable steaming and cooking materials to make Chinese cooking utensils look like "bamboo cooking steamer" which could integrate the traditional Chinese eating habits into advanced modern electronic pot design (Illustration 8-26, Illustration 8-27).

图 8-26 融合传统中国饮食文化的作品《蒸汽加热餐具组合》探索实验图
(原创设计师:郑昱)

第 8 章　工业产品设计
Chapter Eight: Industry Production Design

图 8-27　《蒸汽加热餐具组合》设计作品的材料与造型实验图
（原创设计师：郑昱）

三、设计构思与设计流程（图 8-28、图 8-29）

III. Design Concept and Design Cycle (Illustration 8-28, Illustration 8-29)

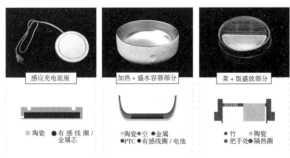

图 8-28　《蒸汽加热餐具组合》设计流程与功能图
（原创设计师：郑昱）

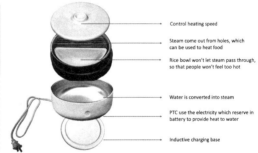

图 8-29　《蒸汽加热餐具组合》设计功能与创意展示图
（原创设计师：郑昱）

四、《蒸汽加热餐具组合》的设计方法与流程（图 8-30~图 8-32）

IV. Design Method and Cycle of the Half & Half (Illustration 8-30, Illustration 8-31, Illustration 8-32)

市场分析 → 客户调研 → 饮食文化调研 → 大类产品调研 → 探索实验 → 材料实验 → 设计定位 → 设计方案 → 模型制作 → 产品展示

Material Analysis → Customer Research → Dietary Culture Survey → Major Product Research → Exploratory Experiment → Material Experiment → Position Design → Design Proposal → Model Building → Work Presentation.

蒸锅功能
用另一片蒸架代替饭碗部分，从而拥有单独的蒸锅功能

煮锅功能
单独使用加热部分，从而拥有煮锅功能（面、汤）

图 8-30　集中国传统菜肴"蒸""煮"烹饪方法于一体的《蒸汽加热餐具组合》设计展示图
（原创设计师：郑昱）

图 8-31　《蒸汽加热餐具组合》设计功能与步骤展示图
（原创设计师：郑昱）

图 8-32　《蒸汽加热餐具组合》设计创意展示图
（原创设计师：郑昱）

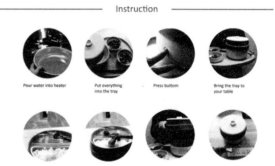

思考题：

1. 请采用双语的方式，谈谈我国古代"造物观念"蕴含的创新思想。
2. 请思考作品《八宝吉祥扇》对你有哪些启发。
3. 请结合作品《延绵·书桌设计》，思考如何在设计中运用传统中国文化元素。

第 9 章　视觉文化设计
Chapter Nine: Visual Culture Design

[本章导读]

近年来,"视觉文化"伴随着设计与互联网、影像与多媒体、UI 界面设计、新媒体版式设计等各种载体已悄然步入人们的日常生活之中,它是集艺术、科学、传媒、技术于一体的跨界设计学科。图像设计是视觉文化设计的基础,它是一种凝练意象的传播语汇,也是一种反映文化内涵的抽象符号,使人们能够直观地认识和理解图像所赋予的设计思想。视觉文化设计具有普及、博识、丰富的特点,通过图像设计等方式传授给读者积极向上的思想观点、文化价值、传统观念,对弘扬传统文化、展现时代精神具有重要的作用。

本章列举了中国设计师俞雪莱的设计作品,采用甲骨文、中国结、中国泥塑等视觉文化元素的设计曾获得国内外专家同行们的好评,这些作品中着重运用了中国传统文化元素,又结合西方文化的多元设计,通过图像设计、设计沟通、设计理念与设计方法的比较,探索"中西文化"

9.1 俞雪莱设计师的心得：设计需要良好的沟通
Feeling of Shirley Yu Designer: Design Requires Good Communication

"设计需要良好的沟通"。[1]俞雪莱毕业于中国美术学院平面设计专业获学士学位，毕业于英国圣马丁艺术学院获设计学硕士学位；之后，她在英国伦敦时装学院、中国东华大学服装设计专业进修。她曾担任多家国际时尚品牌公司、设计公司

Design requires good communication.[1] Yu Xuelai graduated from China Academy of Art and got her Bachelor degree of Visual Communication Design. She graduated from Central Saint Martins, University of Arts London and got her Master Degree of Design. After graduation, she continued her study of Fashion Design in London Fashion Academy in UK and Donghua University in China. She served as the chief designer of several international fashion brand companies and design companies at home and abroad. She served as

图 9-1 设计师：俞雪莱【毕业于中国美术学院获设计学学士学位，毕业于英国圣马丁艺术学院获设计学硕士学位。爱马仕（中国）资深橱窗设计师、如沐展示设计有限公司设计总监。2015年创立设计师品牌"SHIRLEY YU"】

---

[1] 源自俞雪莱设计师访谈录（Records come from an interview with Yu Xuelai designer）。

的首席设计师，国际品牌爱马仕（中国）资深橱窗设计师，如沐展示设计有限公司设计总监。她克服种种艰难，开始在中国创业。2015年，她以自己的名字创立设计师品牌"SHIRLEY YU"，旨在用心追求高端中国品质设计，意在打造都市时尚女性的系列礼服。她是一位跨界设计师，她的设计作品都洋溢着浓浓的中国传统文化气息（图9-1）。

the senior designer of Hermes Paris, the international brand Shop-window design in China. She was the design director of RU MU Design Co., Ltd. She overcame all sorts of difficulties to start her own business in China. As the start-up, she created designer brands "SHIRLEY YU" by her own name in 2015. She is pursuing high-quality Chinese design with great attention and aims to create a serious of dress suits for urban fashion women. She is a cross-boundary designer. Her design works are permeated with a strong and rich flavor of Chinese traditional culture (Illustration 9-1).

9.2 俞雪莱设计师作品实例：甲骨文的设计元素
Work of Shirley Yu Designer: Design Element of Inscription on Bones or Tortoise Shells of the Shang Dynasty

甲骨文是"古代刻在龟甲和兽骨上的文字。现在的汉字就是从甲骨文演变下来的。"[1]甲骨文属于典型的象形文字，也是一种描摹实物形状的文字，或者说是用图形、符号表达含义的文字。设计师俞雪莱运用一个在甲骨上抽象的酒瓶图形来展现中国历史悠久、源远流长的酒文化以及饮食文化（图9-2）。

"This inscription was carved on animal bones or tortoise shells in ancient times. Present Chinese characters were evolved from the inscription on bones or tortoise shells of the Shang Dynasty."[1] This character is not only typical pictograph, but also one kind of script writing that could describe physical shapes, or one kind of script writing that could express meaning in figures and symbols. Shirley Yu Designer applied one abstract bottle of wine figure on tortoise shells to exhibit Chinese long history of wine culture and food culture (Illustration 9-2).

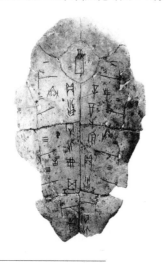

图9-2 视觉文化设计《甲骨文元素》
（原创设计师：俞雪莱）

---

[1] 中国社会科学院语言研究所词典编辑部，《现代汉语词典》，北京：商务印书馆，1985年，第542页。(Dictionary Editorial Department of Chinese Academy of Social Sciences. (1985) Modern Chinese Dictionary. Beijing: The Commercial Press. P542.)

中国的象形文字是华夏民族文化传承的精华，我们的祖先最初从刻画图形、描摹物像来记录日常的生产生活状况。尽管当今已发现有古埃及象形文字、苏美尔象形文字、古印度象形文字等，然而，中国的象形文字被国际学界认为是最为形象达意的汉字字体。设计师俞雪莱设计了一个酒符号代替了甲骨文的一个字，将"象形文字—甲骨文"的中国元素与享誉国际的瑞典伏特加品牌结合运用，体现了"你中有我，我中有你"的设计思想，让世界更了解、更喜爱中国及中国文化（图9-3）。

Chinese pictograph is the essence of Chinese national cultural heritage. Our ancestors initially recorded daily production and living conditions through engraving figures and depicting objects. Ancient Egyptian pictograph, Sumerian pictograph and ancient Indian pictograph have been found today but Chinese pictograph is regarded as the best images to convey thoughts and feelings by international academic circles. Shirley Yu Designer designed to replace the letters on the tortoise shells by the bottle of wine and combined the Chinese element of "pictograph-inscription on Bones or Tortoise Shells" with the Swedish international renowned Absolute Vodka brand together to embody design concept of "you are in my mind and I am in your mind", which could let the world appreciate China and more love Chinese culture (Illustration 9-3).

图9-3 视觉文化设计《甲骨文与瑞典伏特加》
（原创设计师：俞雪莱）

9.3 俞雪莱设计师作品实例：中国结的设计元素
Work of Shirley Yu Designer: Design Element of Chinese Knot

这是为2010年上海世博会设计的招贴海报。此作品中，设计师俞雪莱有意融入具有中国传统文化情结的"中国结"元素，并且选用著名欧洲非物质文化遗产的"手工蕾丝"元素。两者设计于一体，通过象征手法表达视觉文化的寓意，采用"红与黑"两种颜色比喻东方和西方，运用"中国结与西方蕾丝"两种元素象征东、西方的连接与友谊。此次设计的整体组合象征着2010年上海世博会的理念：中西方的紧密结合（图9-4、图9-5）。

This poster was designed for Shanghai, World Expo in 2010. In this work, Shirley Yu Designer consciously tried to use traditional Chinese cultural complex element of Chinese knot and well-known European intangible cultural heritage element of handmade lace in harmony. Both were designed in one to express visual cultural implied meaning through symbolic approach. The Red and the black colours symbolized the west and east; the two elements of Chinese knot and Western lace symbolized the connections and friendship between west and east. The red strip and black lace around represented the meaning of west and east, and the whole combination stands for the world EXPO in 2010 Shanghai: tight integration of west and east(Illustration 9-4, Illustration 9-5).

图9-4 "中国结"元素的2010年世博会海报设计
（原创设计师：俞雪莱）

图9-5 "中国结"元素的海报设计
（原创设计师：俞雪莱）

9.4 俞雪莱设计师作品实例：中国泥塑的设计元素
Work of Shirley Yu Designer: Design Element of Chinese Clay Sculpture

中国泥塑兴起于汉代，经过源源不断的传承与发展，已成为一种深受人们喜爱的民间艺术。泥塑艺术以其独特的三维直观效果和多样化的创作风格，带给人们一种返璞归真的视觉享受。2006年泥塑被列为中国国家级非物质文化遗产。中西方文化交流日益频繁，中国泥塑作为中国最具特色的手工艺品远销海外，享誉世界。

中国泥塑的种类丰富，主要有天津"泥人张"、无锡彩塑、惠山泥塑、凤翔泥塑、敦煌泥塑以及盛行于民间的自创泥塑形式，风格多样、造型丰富、特色鲜明，作为未来设计元素可以挖掘和创作的空间很广阔。设计师俞雪莱以西方童话故事《三只小猪和大灰狼》为创作基础，运用中国泥塑

Chinese Clay Sculpture sprang up all over china since the Han dynasty. After a steady stream of inheritance and development, it became a kind of popular folk art. Clay Sculpture Art could bring people visual enjoyments to recover their original simplicity through its unique three-dimensional visual effects and a variety of creative styles. In 2006, Chinese Clay Sculpture was classified as China's state-level intangible cultural heritage. There were more and more frequent cultural exchange between China and the West. As the one of the most distinctive handicrafts in China, Chinese Clay Sculptures were exported to overseas countries, which would enjoy high reputation in the world.

Chinese Clay Sculptures are rich in variety such as the Clay Figure Zhang in Tianjin, Colored Figure in Wuxi, Huishan Clay Figure, Fengxiang Clay Figure, Dunhuang Clay Figure and self-created forms of clay sculptures prevailing among the people. Chinese Clay Sculptures shows various styles, plentiful modeling, and distinctive characteristics, which would remain very wide spaces to be excavated and created as design elements in the future. Based on the Western

的材料和制作方法，自制了一系列造型别致、惟妙惟肖、生动活泼、充满童趣并且富有中国特色的跨界设计作品（图 9-6~ 图 9-9）。

children's story of Three Little Pigs and A Big Wolf, Shirley Yu Designer utilized materials and making methods of Chinese Clay Sculptures to create a series of absolutely lifelike, vivid, childish and unique modeling works of cross-boundary design with full of Chinese characteristics (Illustration 9-6 to Illustration 9-9) .

图 9-6 "中国泥塑"元素的《三只小猪与狼》跨界设计作品展示图之一
（原创设计师：俞雪莱）

图 9-7 "中国泥塑"元素的《三只小猪与狼》跨界设计作品展示图之二
（原创设计师：俞雪莱）

图 9-8 "中国泥塑"元素的《三只小猪与狼》跨界设计作品展示图之三
（原创设计师：俞雪莱）

图 9-9 "中国泥塑"元素的《三只小猪与狼》跨界设计作品展示图之四
（原创设计师：俞雪莱）

思考题：

1. 请用英文简述"视觉文化设计"的概念。
2. 请用英文谈谈你对作品《甲骨文》的观后感。
3. 请用双语表述的方式，列举视觉设计中蕴含的"中国文化"元素。

第 10 章 历史建筑保护与再利用
Chapter Ten: Historical Architecture Protection and Design

[本章导读]

作为传承人类文明的见证实体，建筑是艺术、设计和科学的完美结合，历史文化建筑是各个时期建筑构造、技术、功能、风格、审美、精神的集中体现，也是科学与艺术的结晶。每一座经典的历史建筑都是一件珍贵的历史文化遗产，它以"古今对话"的文化承载方式，成为人们了解传统文化的一座桥梁，它传达给人们关于时代风貌、文化积淀、历史文脉等多元而广博的史料信息。对历史建筑的保护与再利用设计，一方面使我国优秀的传统文化得以传承，另一方面能够创造出更加富有凝聚力的新时代文化。

本章精选了浙江工业大学艺术学院环境设计系吕勤智教授的《老土木楼建筑保护与再利用》获奖设计项目，这项成果体现出在新旧城市建设发展中，古建筑的保护、改造、更新与再利用是城市发展建设中一项重要任务；现存的历史文化建筑不仅是一个正在失去的时代的记忆和见证，而且本身是一种已经存在的，可以有效利用的资源。吕勤智教授曾说："历史文化建筑的更新与再利用将使老建筑焕发出新的生命力，更是一种对人文资源的继承；它能够带动城市文化的发展，并保持城市、建筑及其所在区域的魅力与活力"。

10.1 吕勤智设计师的体会：设计引领生活
Learning from Lv Qinzhi Designer: Design Eagerly Looks forward to Life

"设计引领生活"。[①]吕勤智教授曾任哈尔滨工业大学建筑学院环境设计学科带头人、环境艺术设计系主任、建筑学院副院长以及浙江工业大学环境设计学科带头人等。现为浙江工业大学小城镇城市化协同创新中心副主任、城市与建筑环境设计研究所所长。2009荣获中国建筑学会室内设计分会"1989-2009中国室内设计二十年学会贡献奖"；获得2010年度教育部全国"宝钢"优秀教师奖；2015年度"中国人居环境设计学年奖"优秀指导教师奖等荣誉。吕勤智教授主持的《老土木楼建筑保护与再利用设计》项目荣获"中国建筑设计奖"银奖（图10-1）。

Design eagerly looks forward to Life.[①] Professor Lv Qinzhi was the academic pacesetter of environmental design department, the director of environmental design department, vice-dean of School of Architecture Harbin Institute of Technology. He was the academic pacesetter of environmental design field in Zhejiang University of Technology. He is not only the vice-director of of Urbanization of Small Towns Collaborative Innovation Center, but also the Institute director of Urban and Architectural Environment Design in Zhejiang University of Technology. Professor Lv Qinzhi was awarded the "Chinese Interior Design Twenty Years Society Contribution Award (1989-2009)" in 2009, the "National Outstanding Teacher Award of Ministry of Education" in 2010, and the "Excellent Guide Teacher of Chinese Living Environment Design Academic Year Award" in 2015. This program titled the Old Civil Construction Protection and Reuse Design presided by Professor Lv Qinzhi was awarded the "Silver Medal of China Architectural Design Award" in 2013 (Illustration 10-1).

---

① 摘自吕勤智教授访谈录。(Records come from an interview with Lv Qinzhi Professor.)

第 10 章 历史建筑保护与再利用
Chapter Ten: Historical Architecture Protection and Design

图 10-1 吕勤智教授（前排左五）《哈尔滨工业大学"老土木楼"建筑保护与再利用设计》项目荣获"中国建筑设计奖"银奖

10.2 吕勤智设计师的作品实例：老土木楼建筑保护与再利用
Work of Lv Qinzhi Designer: Old Civil Construction Protection and Reuse Design

《哈尔滨工业大学"老土木楼"建筑保护与再利用设计——哈尔滨工业大学博物馆室内设计》这项成果基于对历史建筑空间的保护与再生的观念与责任，针对哈尔滨工业大学"老土木楼"建筑的保护与再利用进行理论联系实际的设计研究（图10-2）。

这项设计成果基于对历史建筑空间的保护与再生的理念与原则，从老建筑的历史与文化价值和历史建筑空间再生两个方面进行深入的分析，阐释对具有历史文化建筑的保护、改造与再利用的观点与实施策略；提出针对哈尔滨工业大学"老土木楼"这座历史文化建筑保护与再利用的设计原则与对策，并将其运用于设计与建设实践当中（见图10-3）。

This Achievement "Harbin Institute of Technology Old Civil Construction Protection and Reuse Design—Interior Design of Harbin Institute of Technology Museum" based on the concept and responsibility of protection and regeneration of the historical building space. It was the design research on integrating theory with practice for Old Civil Construction Protection and Reuse of Harbin Institute of Technology (Illustration 10-2).

This Achievement did research work based on the ideas and principles of protection and regeneration of the historical building space. Professor Lv Qinzhi carried on the thorough analysis for two aspects, one was the historical and cultural values of historical building, and the other was the regeneration of historical building space. He interpreted his concept and implementation strategy on protection, reconstruction and regeneration of the historical cultural building. He also proposed his design principles and countermeasures for Old Civil Construction Protection and Reuse of Harbin Institute of Technology and apply to design and construction practice (Illustration 10-3).

第10章 历史建筑保护与再利用
Chapter Ten: Historical Architecture Protection and Design

哈尔滨工业大学博物馆是在保护的前提下,将"老土木楼"建筑的使用功能转换为博物馆的老建筑改造与再利用建设项目,设计中着重强调老建筑本身就是博物馆展品的重要组成部分,也是博物馆最大的展品和镇馆之宝,使建筑本身体现出承载哈尔滨工业大学历史和唤醒历史价值的作用,将体验和展示有机结合在一起(图10-4)。

On the premise of protection, the program of Harbin Institute of Technology Museum converted the use functionality of Old Civil Construction to Protection and Reuse of historical building. In this design, it emphasized that the Old Civil Construction itself must be not only the important component of museum exhibits, but also museum's treasure and the largest exhibit on display. The Old Civil Construction itself was invested with the history of Harbin Institute of Technology, embodied function of awakening historical value, and organically combined experiences and exhibitions (Illustration 10-4).

图10-2 《哈尔滨工业大学"老土木楼"建筑保护与再利用设计》项目展示图之一,百年建筑的历史照片
(设计主持:吕勤智教授;设计团队:吕勤智教授科研团队;项目获奖:"中国建筑设计奖"银奖,2013年)

图10-3 《哈尔滨工业大学"老土木楼"建筑保护与再利用设计》项目展示图之二,百年建筑改造之前的历史照片
(设计主持:吕勤智教授;设计团队:吕勤智教授科研团队;项目获奖:"中国建筑设计奖"银奖,2013年)

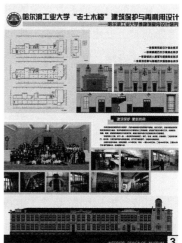

图10-4 《哈尔滨工业大学"老土木楼"建筑保护与再利用设计》项目展示图之三,百年建筑保护、更新与再利用设计方案
(设计主持:吕勤智教授;设计团队:吕勤智教授科研团队;项目获奖:"中国建筑设计奖"银奖,2013年)

10.3 传承文化再创辉煌：获奖设计作品赏析
Inheriting Culture and Splendid Civilization: Appreciation of Prize-winning Design Work

新建的哈尔滨工业大学博物馆注重强调对历史建筑的保护，挖掘老建筑的文化内涵，赋予这座老房子新的使用价值。老房子被赋予新的生命和价值，成为展示哈工大文化，承载哈工大精神，见证哈尔滨工业大学发展历史的大学博物馆（图10-5）。

这座"老土木楼"始建于1906年，是一座历经百年沧桑的历史建筑。该项目在设计中努力从文化和技术两方面最大限度地实现保护与再利用的目的，运用传统与现代相结合的手法，将此建筑更新改造为传播哈工大文化和精神的博物馆，成为一座被用心设计的老房子，一座展示大学文化与精神的老房子，一座感悟历史和启迪未来的老房子（图10-6）。

Newly-built Harbin Institute of Technology Museum emphasizes the protection and cultural connotation of historical building. This Old Civil Construction had not only entrusted new value but also new life. This Old Civil Construction became the university museum, which would exhibit culture of Harbin Institute of Technology, bear spirit of Harbin Institute of Technology, and witness the historical development of Harbin Institute of Technology (Illustration 10-5).

This Old Civil Construction was founded in 1906. It was a historical building that experienced many vicissitudes more than one hundred years. Professor Lv and his design team made great efforts to realize the purpose of protection and reuse maximumly from two aspects such as culture and technology and applied the combining of traditional and modern methods. The Old Civil Construction had been transformed into a museum that could spread culture and spirit of the Harbin Institute of Technology. It became an attentively designed old building, which would exhibit culture and spirit of university to understand history and enlighten the future (Illustration 10-6).

第 10 章 历史建筑保护与再利用
Chapter Ten: Historical Architecture Protection and Design

设计与建设中努力保护和留存建筑中的历史遗迹，展现老建筑的沧桑与厚重的历史感，让建筑本身述说故事，留给后人可品味与体验、重温与遐想的空间环境。老房子使人能够找回历史记忆，感悟历史和启迪未来，成为学校历史发展的见证和城市历史文化的组成部分（图10-7）。

In this process of designing and building, professor Lv and his design team tried hard to protect and preserve historical traces and monuments in architecture, exhibit historical sense of the Old Civil Construction with vicissitudes and messiness and tell the story on this building itself, which could create environmental space of tasting, experiencing, reviewing and revering for later generations. The Old Civil Construction became testimony of university historical development and integrant part of urban historical culture, which could enable people to retrieve historical memories, understand history and enlighten the future (Illustration 10-7).

图 10-5 《哈尔滨工业大学"老土木楼"建筑保护与再利用设计》项目展示图之四, 百年建筑保护、更新与再利用设计方案
（设计主持：吕勤智教授；设计团队：吕勤智教授科研团队；项目获奖："中国建筑设计奖"银奖，2013年）

图 10-6 《哈尔滨工业大学"老土木楼"建筑保护与再利用设计》项目展示图之五, 百年建筑保护、更新与再利用设计方案
（设计主持：吕勤智教授；设计团队：吕勤智教授科研团队；项目获奖："中国建筑设计奖"银奖，2013年）

图 10-7 《哈尔滨工业大学"老土木楼"建筑保护与再利用设计》项目展示图之六, 百年建筑保护、更新与再利用设计方案
（设计主持：吕勤智教授；设计团队：吕勤智教授科研团队；项目获奖："中国建筑设计奖"银奖，2013年）

项目获奖时间及级别：

2013年，荣获中国建筑设计最高奖"中国建筑设计奖"银奖。

Silver Medal of China Architectural Design Award, the Top Prize of Chinese Architectural Design in 2013.

获奖作者：

设计主持：吕勤智教授

Design Direct: Professor Lv Qinzhi

设计团队：吕勤智教授科研团队

Design: Professor Lv Qinzhi Design and Research Team

作品出处：

吕勤智，于稚男，王一涵．历史建筑保护与再利用．北京：中国建筑工业出版社，2013．

Lv Qinzhi, Yu Zhinan and Wang Yihan. Protection and Reuse of Historical Architecture. Beijing: China Architecture & Building Press, 2013.

思考题：

1. 请思考《老土木楼建筑保护与再利用》设计项目对你有哪些启发。
2. 请采用双语方式表述古建筑保护与再利用设计的意义。
3. 建议进行一次古建筑考察，用心收集考察资料，撰写一篇古建筑改造设计实习报告。

附录一：中国古典设计文献分类辑录
The First Appendix: The Compilation of Chinese Classical Design Texts

笔者围绕"设计文化"的要义对现存古籍文献进行较为详细的梳理与分类。第一类：中国古代设计文化思想文献，收录22种；第二类：展现中国古代设计文明成就文献辑录，收录11种；第三类：阐释中国古代设计营造方法文献辑录，收录16种。

一、蕴含中国古代设计文化思想文献辑录

[1]《周易集解略例》一卷。（魏）王弼撰，（唐）刑璹注。书中很多处基于古人的文化传统与社会活动，提出我国早期的设计理论思想。见明崇祯三年（1621年），虞山毛汲古阁刻本、上海博古斋影印本，1922年版。或见中华书局铅印本，1985年版。

[2]《韩非子》二十卷。（战国）韩非撰。书中阐释了我国古代造物工艺的原创思想，反映了当时民俗文化对造物的影响。见明万历十年（1582年）刻本、清光绪元年（1875年）湖北崇文书局刻本。或见（清）王先慎撰，上海：上海古籍出版社，1995年版，影印本。

[3]《吕氏春秋》二十六卷。（秦）吕不韦撰。书中借助造物现象与文化观念来阐释当时的治国见解，成为记载古代工艺设计理念的史料文献之一。见元至正年间（1341~1368年）嘉兴路儒学刻本、明末（1621~1644年）朱梦龙刻本。或见（汉）高诱注，北京：北京图书馆出版社，2005年版，影印本。

[4]《洛阳名园记》一卷。（北宋）李格非撰。书中记录了中国古典文化熏陶下古代文人雅士的私家园林设计及其构筑风范。见明崇祯年间（1628~1644年）毛氏汲古阁刻本、清嘉庆十年（1805年）虞山张氏旷照阁刻本。或见影印本，台湾：台湾商务印书馆，1984年版。

[5]《砚笺》四卷。（宋）高似孙撰。砚是中国文房四宝之一，为传播中国文化发挥重要作用，也蕴藏着中国文化意象，书中记述了砚的种类、形态、审美及其与古代文人的渊源奥妙。见清康熙四十五年（1706年）扬州诗局刻本。或见影印本，上海：上海古书流通处，1921年版。

[6]《古玉图谱》一百卷。（宋）龙大渊撰。书中分国宝部、文房部、陈设部等九大部分详细地记载了古代玉器的形态、分类、用途和工艺，诸如文房部的如意、水丞均配有图解，体现古代物质文化与精神文化的融合统一。见清乾隆四十四年（1779年）刻本。或见影印本,济南:齐鲁书社,1995年版。

[7]《洞天清禄》一卷。(宋)赵希鹄撰。书中列举古琴辨、古钟鼎彝辨、笔格辨、古画辨、古砚辨等体现了我国古典审美文化,并阐述了古代工艺理论与历史沿革。见清嘉庆四年(1799年)桐川顾氏刻本、清道光二十九年(1849年)刻本。或见铅印本,中华书局,1985年版。

[8]《辍耕录》三十卷。(元)陶宗仪撰。此书以杂记的形式记述了当时人们的文化生活,从传统雕刻、陶瓷、染织等反映工艺美术特色。见明崇祯三年(1621年)虞山毛氏汲古阁刻本、清光绪十一年(1885年)上海福瀛书局刻本。或见影印本,上海:上海古籍出版社,2012年版。

[9]《宣德鼎彝谱》八卷。(明)吕震撰。此书论及明宣德鼎彝的造型、色彩和铸造等,宣德鼎是中国古代器物文化的典范之一,见清光绪九年(1883年)刻本、影印本,上海:博古斋,1921年版。或见铅印本,中华书局,1985年版。

[10]《长物志》十二卷。(明)文震亨撰。此书分为12个种类,阐释了园之营造和物之陈设,反映了古人物质文化的鉴赏力与创造力。见1984年台湾新文丰出版公司影印本。或见2012年北京中华书局影印本。

[11]《方氏墨谱》六卷。(明)方于鲁撰。书中论及我国传统笔墨文化,古代墨锭的造型与纹样设计。见明万历年间(1573~1620年)方氏美荫堂刻本。或见影印本,上海古籍出版社,1995年版。

[12]《清秘藏》二卷。(明)张应文撰。此书是我国古代工艺美术鉴赏著作,琴棋书画中蕴含中国文化的精华。见清(1644~1911年)抄本、清同治十年(1871年)古冈刘氏藏修书屋刻本。或见影印本,台湾商务印书馆,1984年版。

[13]《秋园杂佩》一卷。(清)陈贞慧撰。此书以笔记的形式论及书砚、折扇、器物形制等,凝练了古人造物文化观念。见清道光十三年(1833年)世楷堂刻本、清咸丰三年(1853年)南海伍氏刻本。或见影印本,台北:新兴书局有限公司,1981年版。

[14]《扬州画舫录》十八卷。(清)李斗撰。此书以笔记的形式记载了古代扬州地区的社会文化,特别是古典园林布局、设计与造景。见清乾隆六十年(1795年)刻本、1931年中国营造学社铅印本。或见影印本,上海古籍出版社,1995年版。

[15]《观石录》一卷。(清)高兆撰。书中论及相石、解石及其创作经验,石雕艺术再现中国古典文化的造物观念。见清康熙三十四年(1695年)新安张氏霞举堂刻本、1936年上海神州国光社铅印本。或见影印本,上海博古斋,1920年版。

[16]《石谱》一卷。(清)诸九鼎撰。书中记载了我国古代用石造景的文化传统。见1920年上海博古斋影印本、1936年上海神州国光社铅印本。或见1985年中华书局铅印本。

[17]《西清砚谱》二十四卷。(清)于敏中撰。书中图文并茂地收录了皇家所收藏的各类古砚,工笔手绘古砚的各种造型设计,展现古人工艺设计的成就,对中国文化的传承与发展产生重要作用。见1934~1935年上海商务印书馆影印本。或见上海书店,1991年版。

[18]《阳羡名陶录》二卷。(清)吴骞撰。此书上卷论及制陶的选材、工艺和制陶家,下卷阐述

古代制陶文化与陶器发展史。见清道光十三年（1833年）吴江沈氏世楷堂刻本、清光绪十五年（1889年）仁和许增娱园刻本。或见影印本，上海古籍出版社，1995年版。

[19]《博物要览》十二卷。（清）谷应泰撰。书中记述古鼎彝、铜器、瓷器、奇石、锦缎等工艺、色彩、形制及出产，反映我国博古藏宝文化的资料集。见清嘉庆十四年（1809年）刻本、清道光五年（1825年）李朝夔刻本。或见影印本，济南齐鲁书社，1995年版。

[20]《履园丛话》二十四卷。（清）钱泳撰。书中以笔记形式论及装潢、雕刻、营造等古代工艺制作，体现中国古代文化诸方面状况。见清道光五年（1825年）述德堂刻本、1979年北京中华书局刻本。或见影印本，上海古籍出版社，1995年版。

[21]《竹人录》二卷。（清）金元钰撰。书中收录了历代竹刻工艺及其沿革，体现我国古人的生态文化观。见1922年嘉定铅印本、1936年上海神州国光社铅印本。或见浙江人民美术出版社，2016年版。

[22]《闲情偶记》十六卷。（清）李渔撰。此书以杂著的形式体现古代中国文化，其"居室部"与"器玩部"记载我国民居建筑的室内设计理论以及园林文化要论。见清康熙十年（1671年）刻本、1936年上海杂志公司铅印本。或见影印本，湖北人民出版社，2002年版。

二、展现中国古代设计文明成就文献辑录

[1]《墨子》十五卷，目录一卷。（战国）墨翟撰。书中基于工艺实践总结出古代制造器物的手工艺经验，也是造物工艺的智慧结晶。见清乾隆四十九年（1784年）灵岩山馆刻本、清光绪元年（1875年）湖北崇文书局刻本。或见北京中华书局，2017年版。

[2]《梦溪笔谈》二十六卷、补笔谈三卷、续笔谈一卷。（北宋）沈括撰。此书内容包括天文、地理、农业、建筑、水利、文学、艺术等诸多领域，展现了古代造物设计的突出成就，成为一部反映我国古代文明史料典籍。见明崇祯年间（1628~1644年）毛氏汲古阁刻本、石印本，上海：上海文明书局，1915年版。或见铅印本，中华书局，1985年版。

[3]《宣和博古图》三十卷。（北宋）王黼撰。此书全面系统地阐述了宋代青铜器的精华，包括器物形制、图像、释文、色彩、尺度等各个方面，是我国古代设计文明的集中体现。见明万历泊如斋刻本、明万历二十七年（1599年）于承祖刻本。或见影印本，北京：北京图书馆出版社，2005年版。

[4]《砚史》一卷。（宋）米芾撰。书中记载了鉴定砚石优劣的方法，列举了晋砚、唐砚至宋砚的造型变化、不同材质砚石之特点及其沿革，体现了我国古代的造物设计文明发展。见明末（1621~1644年）刻本、清嘉庆十年（1805年）虞山张氏旷照阁刻本。或见影印本，中华书局，1985年版。

[5]《格古要论》三卷。（明）曹昭撰。此书是我国最早的文物鉴定著作，其中记录的珍奇工艺与造物观念体现了我国古代文明的智慧结晶。见上海：涵芬楼影印本，1940年版、上海：上海古籍出版社，1995年版。或见北京：中华书局，2012年版。

[6]《游具雅编》。（明）屠隆撰。书中记载了古人发明便于游览的工具，图文并茂地展现我国古人造物设计的才华。见上海涵芬楼影印本，1920年版。或见济南齐鲁书社影印本，1995年版。

[7]《西清古鉴》四十卷。（清）梁诗正撰。书中论及鼎、尊、彝等古代青铜器造型、分类与纹饰，这是一部青铜文化的典籍，也是我国古代文明的文化载体。见清光绪十四年（1888年）上海鸿文书局石印本。或见石印本，上海云华居庐，1926年版。

[8]《匋雅》二卷。（清）寂园叟撰。书中论及古陶起源、历代名窑、匠作、装饰、瓷器样式及其制作工艺等，阐释了瓷器的古代文明。见1918年聚珍仿宋印书局铅印本、1923年上海元昌书局石印本。或见山东画报出版社，2010年版。

[9]《古铜瓷器考》二卷。（清）梁同书撰。书中考证了自唐宋以来官窑及其瓷器造物设计，展现文明昌盛的中国形象，对古代瓷器设计、辨伪、鉴赏具有开创之功。见1936年上海神州国光社铅印本。或见1947年刻本。

[10]《笔史》一卷。（清）梁同书撰。书中论及毛笔的工艺与形制，蕴藏古典文化内涵，体现中华文明的魅力。见清光绪十五年（1889年）仁和许增娱园刻本、1915年上海广益书局刻本。或见铅印本，中华书局，1985年版。

[11]《陶说》六卷。（清）朱琰撰。此书论及陶器起源、历代制陶工艺，是一部古代陶瓷工艺文明史。见清嘉庆元年（1796年）石门马氏大酉山房刻本、1915年上海文明书局石印本。或见山东画报出版社，2010年版。

三、阐释中国古代设计营造方法文献辑录

[1]《考工记》二卷。作者不详。书中记载自先秦以来制车、礼器、钟磬、建筑等六大类的营造方法，体现了古人的造物成就及其唯物求新的思想。见（唐）杜牧注，清光绪十三年（1887年）会稽董氏云瑞楼，木活字本、明万历（1368~1644年）吴兴闵齐伋刻本。或见影印本，（清）王宗涑撰，上海：上海古籍出版社，1995年版。

[2]《水经注》四十卷。（北魏）郦道元撰。此书记载古代关于城邑营造、建置沿革的状况，旁征博引、追本溯源，体现我国古代建造设计领域的成就。见清乾隆四十二年（1777年）黄氏槐荫草堂刻本、清光绪三年（1877年）湖北崇文书局刻本。或见影印本，北京：北京图书馆出版社，2002年版。

[3]《梓人传》。（唐）柳宗元撰。书中阐述了营造设计与工艺创作中总体设计的准则和方法。见《河东先生全集录》六卷，清康熙四十四年（1705年）松鳞堂刻本。或见《河东先生文集》六卷，清宣统二年（1910年）上海会文堂石印本。

[4]《营造法式》三十六卷。（北宋）李诫撰。书中阐述了古代建造技术、建筑设计、施工方法及其营造经验总结，这是一部体系完备、有章可循的古代建筑典籍，对现代建筑发展颇具启迪意义。见

1925年刻本、1933年上海：商务印书馆，铅印本。或见影印本，北京：中国书店，1995年版。

[5]《燕几图》一卷。（宋）黄伯思撰。书中阐述了古代家具制作方法，列举组合式桌子的格局，图文并置、设计巧妙。见明万历年间（1573~1620年）茅一相刻本、清嘉庆十四年（1809年）姚椿抄本。或见影印本，上海：上海科学技术出版社，1984年版。

[6]《蜀锦谱》一卷。（元）费著撰。书中详述了蜀锦图案技术与方法。见明末（1621~1644年）刻本、清顺治三年（1646年）李际期宛委山堂刻本。或见铅印本，中华书局，1985年版。

[7]《古玉图》二卷。（元）朱德润撰。此书记载了璧、带、钩、瑱、珑等玉器的形态与制造工艺。见清乾隆十七年（1752年）亦政堂刻本、清顺治年间两浙督学周南李际期宛委山堂刻本。

[8]《天工开物》三卷。（明）宋应星撰。此书是记载我国古代农业和手工业的生产技术和工艺装备的综合性著作，涵盖陶瓷、车船、金属、农具等的造物方法，被西方学者誉为"17世纪中国工艺百科全书"。见影印本，上海：华通书局，1930年版。或见影印本，北京：国际文化出版公司，1995年版。

[9]《装潢志》一卷。（明）周嘉胄撰。此书论述我国古代书画装裱工艺的步骤、技术和方法。见清康熙三十六年（1697年）刻本、清康熙四十五年（1706年）扬州诗局刻本、清同治十年（1871年）古冈刘氏藏修书屋刻本。或见影印本，上海：上海古籍出版社，1995年版。

[10]《纸墨笔砚笺》一卷。（明）屠隆撰。书中详细阐述了宋纸、元纸、国朝纸等古代造纸方法及其特征。见铅印本，上海：神州国光社，1936年版。或见1947年刻本。

[11]《园冶》三卷。（明）计成撰。书中阐述我国古典园林营造原理和方法，总结了古代造园设计理论。见影印本，上海古籍出版社，1995年版。或见浙江人民美术出版社，2013年版。

[12]《端溪砚史》三卷。（清）吴兰修撰。书中从石质选材到琢砚诸法，记述了中国古代制砚的技艺与制法。见清咸丰九年（1859年）古歙叶砚农刻本、清光绪十五年（1889年）仁和许增娱园刻本。或见铅印本，中华书局，1985年版。

[13]《绣谱》一卷。（清）丁佩撰。书中详细介绍了刺绣的选材要求、工艺步骤、色彩特点、创作方法与优劣品评。见1936年上海神州国光社石印本。或见影印本，上海古籍出版社，1995年版。

[14]《工段营造录》。（清）李斗撰。书中记述了清代营造标准、建造方法与工程法则。见1931年中国营造学社铅印本。

[15]《南窑笔记》一卷。（清）张九钺撰。书中准确记录了景德镇鼎盛时期瓷器生产过程与制作方法，对我国陶瓷制作及发展具有重要史料价值。见1936年上海神州国光社铅印本。或见广西师范大学出版社，2012年版。

[16]《琉璃志》一卷。（清）孙廷铨撰。书中阐释了呈色、火候、配色等琉璃制造过程与造型方法。见清道光十三年（1849年）世楷堂刻本、清道光二十九年（1849年）吴江沈氏世楷堂刻本。或见铅印本，上海神州国光社，1936年版。

附录二：西方艺术设计文献分类辑录
The Second Appendix: The Compilation of Western Art Design Texts

西方艺术设计书籍丰富多彩，从设计教育到设计理论、设计文化，各类设计作品集、设计师个人专辑等举不胜举。笔者精选20部反映西方设计文化、设计思想、设计教育成果并具有代表意义的文献资料，可作为读者自学之时深入研究、启迪思维、设计创作的参考资料。

一、西方设计教育文献辑录

[1] Dow, Arthur Weslay.(1920) *Composition*. (9th ed.) 翻译为《构成》，此书由阿瑟·韦斯利·道撰写，最早于1899年初版，直至1940年期间再版19次，成为在西方影响广泛而深远的美术教材。

[2] Victor D'Amico (1942) *Creative Teaching in Art*. 翻译为《创造性艺术教学》，这本书的作者维克托·达米科对于20世纪30年代的西方艺术教育产生重要影响，该书以不同艺术创作材料，诸如陶艺、雕塑、绘画、书法、拼贴等分各个章节来论述，同时书中强调培养学生创造力的重要作用。

[3] 德国包豪斯曾使用的教科书：Gyorgy Kepes (1944) *Language of Vision*. 翻译为《视觉语言》，这部教科书特别提出了以设计基础知识、基础设计能力来取代传统西方美术学校中的素描等基础课程的改革理念。

[4] 德国包豪斯曾使用的教科书：Moholy Nagy (1947) *Vision in Motion*. 翻译为《运动中的视觉》，这部教科书由莫霍利·纳吉编写，内容包括关于包豪斯基础教学的探索及课程研究专案。

[5] Smith, Walter (1872) *Art Education Scholastic and Industrial*. 翻译为《学校艺术教育与工业艺术教育》，著者沃特·史密斯于1872年写成的一部教材，阐释装饰主题的各种徒手绘画法，包涵了生产与工业艺术在内的所有图画门类。由美国波士顿的詹姆斯·奥斯古德出版公司1872年出版。

[6] Journal：Applied Arts Book. (1901) 翻译为《应用艺术手册》，后更名为 The School Art Books. 翻译为《学校艺术指南》，这本书阐释了自然写生画、美的鉴赏、实物画、模型画等艺术教学课程指导内容及方法。此书成为提供给美术教师的关于艺术活动种类的参考资料。

二、西方设计文化与设计理论文献辑录

[1] Marcus Vitruvius Pollio. (1914) Ten Books on Architecture. 翻译为《建筑十书》,这本书是西方传承至今最早的一部建筑设计理论书,书中总结了古代城市规划、建筑设计、设计美学等基本原理。由美国哈佛大学出版社 1914 年出版。

[2] Halsey, A. H. (Ed.). (1997) EDUCATION: Culture, Economy, and Society. 翻译为《教育:文化、经济和社会》,这本书由六大部分的 52 篇论文组成,主要论述了文化是可见与不可见的因素,文化创新对社会发展的作用,在后现代主义时期文化的影响力等内容。由英国牛津大学出版社 1997 年出版。

[3] Freeman, Kerry. (2003) Teaching Visual Culture Curriculum, Aesthetics, and the Social Life of Art. 翻译为《视觉文化教育课程:美学与艺术中的生活》,这本书的写作思路广阔,包括艺术史论、设计理论和设计评论等领域,重点论述了视觉文化与创造力的内在联系。由美国纽约师范大学出版社 2003 年出版。

[4] Raizman, D.(2003) History of Modern Design. 翻译为《现代设计史》,这本书分为六个部分共 16 篇章,阐释从 1700 年设计需求至 2010 年设计多元化发展状况,包罗万象地论述了西方设计沿革与文化脉络以及设计案例中的文化元素。由英国劳伦斯王者出版公司 2003 年出版。

[5] Charles A. Birnbaum, Mary V. Hughes. (Ed.). (2005) Design with Culture: Claiming America's Landscape Heritage. 翻译为《设计与文化:美国景观文化遗产》,这本书从设计与文化的角度论述了历史建筑与人文景观保护对发展旅游业、解决生态问题的具体举措和重大意义,同时探讨文化传统与景观保护的起源。由美国弗吉尼亚大学出版社 2005 年出版。

[6] Julier, Guy. (2007) The Culture of Design. 翻译为《设计文化》,这本书不仅论述设计文化理论,还有详细的设计案例分析;书中阐述了品牌设计与文化、设计产品与文化、消费与文化等精辟观点以及设计师的心得体会。由赛奇国际出版公司 2007 年出版。

[7] Lidwell, W., Holden, K. and Butler, J. (2010) Universal Principles of Design. 翻译为《设计普适原则》,这本书强调设计是一门跨学科的专业,阐释如何将视觉文化概念运用于设计实践,提出各种设计参考标准和设计技能要求。

[8] Celi, Manuela. (Ed.) (2015) Advanced Design Cultures: Long-Term Perspective and Continuous Innovation. 翻译为《先进的设计文化:长远而持久的创新力》,这本书分为两大部分的 11 个篇章论述了文化对于先进设计的重要性,唯有尊重文化与历史才能更好地规划未来,也阐释了关于产品设计的革新思想。由斯普林格国际出版公司 2015 年初版。

三、西方设计作品专辑

[1] Cruickshank, D.(1998) Architecture: A Critics Choice. 翻译为《经典建筑专辑》,这本书分为十章

介绍西方建筑发展史以及著名建筑，图文并茂地再现了西方建筑风格的流变、建筑材质与建筑方法的进程。由美国艾薇出版公司1998年出版。

[2] Smith, Edwin.（1999）Architecture in Britain and Ireland 600-1500. 翻译为《英国与爱尔兰的建筑：600-1500》，这本书图文并茂地展现了兴建于600~1500年英国与爱尔兰的历史建筑。由伦敦哈维尔出版社1999年出版。

[3] Stephenson, David. (2005) Visions of Heaven: The Dome in European Architecture. 翻译为《视觉的天堂:欧洲建筑中的穹顶》，这本书根据历史主线，展现了西方从古典主义建筑、古罗马建筑、古希腊建筑、哥特式建筑、文艺复兴建筑、巴洛克与洛可可建筑到19世纪建筑的穹顶设计，穹顶装饰展现了油画、雕塑、壁画的艺术成就，是一部集历史文化、艺术与建筑于一体的设计专辑。由纽约普林斯顿建筑出版社2005年出版。

[4] Munari, Bruno. (2009) Design as Art. 翻译为《艺术的设计》，这本书的作者是一位著名设计师，书中辑录了很多精彩的设计实例。作者以设计师的视角阐释了关于视觉文化设计、图形设计、工业设计的趣味性、启发性以及实用价值。由英国企鹅出版集团公司2009年出版。

[5] Milton, A.and Rodgers, P. (2011) Product Design. 翻译为《产品设计》，这本书一方面介绍了杰出的设计师，一方面阐述了设计师的设计作品，提出产品设计中的文化引领，与众不同的设计方法与设计思维。由英国劳伦斯王者出版公司2011年出版。

[6] Alegre, I. (2013) Start Product Designers: Prototypes, Products, and Sketches from the World's Top Designers. 翻译为《产品设计师启蒙：世界顶尖设计师的草图、模型和产品设计》，这本书展示当代顶尖级设计师的创意产品，收集了设计师的产品设计作品、内涵丰富、充满创意的构思草图及其模型350幅全彩图，同时，论述了如何高效地做出美观、实用而迎合消费者的产品设计。由哈珀设计出版公司2013年出版。

征引及参考文献
References

（一）古籍文献

[1]（周）卜商.《子夏易传》卷一.见清通志堂经解本.

[2]（宋）蔡沈.《书经集传》.见清文渊阁四库全书本.

[3]（宋）郭熙.《林泉高致》.见明代读书坊刻本.

[4]（宋）李格非.《洛阳名园记》.见明崇祯毛氏汲古阁刻本.

[5]（宋）黄伯思.《燕几图》一卷.见明万历茅一相刻本.

[6]（宋）李诚.《营造法式》三十六卷.见上海商务印书馆铅印本.

[7]（宋）赵希鹄.《洞天清禄》一卷.见清嘉庆四年桐川顾氏刻本.

[8]（元）佚名.《大元毡罽工物记·御用》.见上海仓圣明智大学铅印本.

[9]（元）陶宗仪.《辍耕录》三十卷.见明崇祯三年虞山毛氏汲古阁刻本.

[10]（明）宋应星.《天工开物》.见上海华通书局影印本.

[11]（明）文震亨.《长物志》十二卷.见北京中华书局影印本.

[12]（明）计成.《园冶》三卷.见中国营造学社铅印本.

[13]（明）周嘉胄.《装潢志》一卷.见清康熙四十五年扬州诗局刻本.

[14]（明）方于鲁.《方氏墨谱》六卷.见明万历方氏美荫堂刻本.

[15]（清）李斗.《工段营造录》.见中国营造学社铅印本.

[16]（清）谷应泰.《博物要览》十二卷.见清道光五年（1825年）李朝夔刻本.

[17]（清）寂园叟.《匋雅》.见静园刻本.

[18]（清）金元钰.《竹人录》二卷.见上海神州国光社铅印本.

[19]（清）李渔.《闲情偶寄·器玩部》.见清康熙十年刻本.

[20]（清）钱泳.《履园丛话·造园》.见清道光五年述德堂刻本.

(二)中文文献

[1] 蔡元培.大学的意义:蔡元培卷.济南:山东文艺出版社,2006.

[2] 蔡元培.文明的呼唤:蔡元培文选.天津:百花文艺出版社,2002.

[3] 曹意强.二十世纪的中国绘画.杭州:浙江人民美术出版社,1997.

[4] 陈瑞林.20世纪中国美术教育历史研究.北京:清华大学出版社,2006.

[5] 董占军,郭睿.外国设计艺术文献选编.济南:山东教育出版社,2012.

[6] (美)大卫·瑞兹曼著.若斓达·昂(澳).现代设计史(第二版)李昶译.北京:中国人民大学出版社,2013.

[7] 杭间.设计的善意.桂林:广西师范大学出版社,2011.

[8] 何明,廖国强.中国竹文化.北京:人民出版社,2007.

[9] 吕勤智,于稚男,王一涵.历史建筑保护与再利用.北京:中国建筑工业出版社,2013.

[10] 倪镔.智设计·活文化.北京:清华大学出版社,2015.

[11] 倪建林,张抒.中国工艺文献选编.济南:山东教育出版社,2012.

[12] 卢世主.从图案到设计:20世纪中国设计艺术史研究.江西人民出版社,2011.

[13] 潘耀昌.20世纪中国美术教育.上海:上海书画出版社,1999.

[14] 孙德明.中国传统文化与当代设计.北京:社会科学文献出版社,2015.

[15] 沙里宁·伊利尔(美).形式的探索:一条处理艺术问题的基本途径.北京:中国建筑工业出版社,1989.

[16] 邱春林.设计与文化.重庆:重庆大学出版社,2009.

[17] 尹定邦.设计学概论.长沙:湖南科学技术出版社,2008.

[18] 应宜文.美国《视觉艺术教育法规》述评.美术研究,2014,11(156).

[19] 应宜文.论文学家鲁迅的《拟播布美术意见书》.中文学术前沿,2012,5(4).

[20] 张旅平.多元文化模式与文化张力.北京:社会科学文献出版社,2014.

[21] 曾繁仁.现代中西高校公共艺术教育比较研究.北京:经济科学出版社,2009.

[22] 祝帅.中国文化与中国设计十讲.北京:中国电力出版社,2008.

[23] 中华人民共和国文化部教育科技司编.中国高等艺术院校简史集.杭州:浙江美术学院出版社,1991.

[24] 紫图大师图典编辑部.新艺术运动大师图典.西安:陕西师范大学出版社,2003.

（三）外文图书

[1] Anfam, David. Abstract Expressionism. London: Thames and Hudson, 1999.

[2] Biggs, John. Teaching for Quality Learning at University. Philadelphia: SRHE & Open University Press, 1999.

[3] Bruner, Jerome. The Process of Education. U.S.A. Cambridge: Harvard University Press, 1960.

[4] Clapp, Edward P. (Ed.). 20 UNDER 40: Re-Inventing the Arts and Arts Education for the 21st Century. Bloomington: Author House, 2011.

[5] Celi, Manuela. (Ed.). Advanced Design Cultures: Long-Term Perspective and Continuous Innovation. Milano: Springer International Publishing, 2015.

[6] D' Amico, Victor. Creative Teaching in Art International Textbooks in Art Education. Scranton: International Textbook Company, 1942.

[7] Davidson, Susan. (Ed.). Art in America Three Hundred Years of Innovation. London: Merrell Publishers Limited, 2007.

[8] Donald, James. Sentimental Education: Schooling, Popular Culture and the Regulation of Liberty. London: VERSO, 1992.

[9] Dow, Arthur Weslay. Composition(9th ed.). New York: Doubleday, Page and Company, 1920.

[10] Efland, Arthur D. A History of Art Education: Intellectual and Social Currents in Teaching the Visual Arts. New York: Teachers College Press, 1990.

[11] Halsey, A. H. (Ed.). EDUCATION: Culture, Economy, and Society. Oxford: Oxford University Press, 1997.

[12] Hetland, Lois. Studio Thinking: the Real Benefits of Visual Arts Education. New York: Teachers College Press, 2007.

[13] Fishel, Catharine. How to Grow as a Graphic Designer. New York: Allworth Press, 2005.

[14] Freeman, Kerry. Teaching Visual Culture Curriculum, Aesthetics, and the Social Life of Art. New York: Teachers College Press, 2003.

[15] Gardner, Howard. Art Education and Human Development. Los Angeles: Getty Publications, 1990.

[16] Lee, Alison. & Danby, Susan. (Ed.). Reshaping Doctoral Education International approaches and pedagogies. London: Routledge, 2012.

[17] LUCIE-SMITH, Edward. Lives of the Great Modern Artists. London: Thames & Hudson, 1999.

[18] Marcus Vitruvius Pollio. Ten Books on Architecture. Harvard University Press, 1914.

[19] McLanathan, Richard. Art in America A Brief History. New York: Harcourt Brace Jovanovich, Publishers, 1973.

[20] Moore, Alex. Teaching and Learning: Pedagogy, Curriculum and Culture. London: Routledge, 2001.

[21] National Art Education Association. Design Standards for School Art Facilities. Publisher: National Art Education Association, 1993.

[22] Noblit, George W. & Corbett, H. Dickson. Creating and Sustaining Arts-Based School Reform. New York: Routledge, Taylor & Francis Group, 2009.

[23] Prather, Marla F. & Arnason, H. H. History of Modern Art. New York: Harry N. Abrams, Inc., Publishers, 1998.

[24] Proefriedt, William A. High Expectations: The Cultural Roots of Standards Reform in American Education. New York: Teachers College Press, 2008.

[25] Smith, Edwin. Architecture in Britain and Ireland 600-1500. London: The Harvill Press, 1999.

[26] Stephenson, David. Visions of Heaven: The Dome in European Architecture. New York: Princeton Architectural Press, 2005.

后 记
Postscript

  21世纪我国高校设计艺术专业人才的培养不仅需要掌握扎实的文化功底与设计基础，更需要将优秀的中国传统文化融入当代设计，将工匠精神与现代设计相结合，弘扬中国传统文化传承与再创新。中国设计文化尚处于初步发展阶段，今后有待探宝挖掘的空间广阔，我们坚信它会有枝繁叶茂的丰硕季节，更将有回馈社会的实际成果。将我国优秀的民族文化作为基石，有益于设计师们拓展国际化视野，有益于促进国际交流并展示我国设计水平，在未来全球化设计的趋势下，涌现更多的中国原创设计佳作。

  2012年，笔者在发表的《高校设计艺术专业双语教学改革的思考》一文中提出："双语教学改革所面临的困难，一是理解设计作品不仅在'外观'，更在于了解设计'文化与审美'的本源；二是适合我国高校设计艺术专业双语教学的教材也面临扩充与改革。在未能引进国际领先的原版教材或选用专业对口的双语教材，又未能建立完善的双语教学评价体系状况下，确实给专业教师开展双语教学带来诸多困难"。近年来，笔者一直自我奋蹄、默默耕耘、尽心尽力地撰写双语教材，笔者希望借此对设计学双语教学作出一份贡献，更希望借此引起设计界对中国文化与设计理论问题的关注，本教程为建立中国自己的设计文化理论打好基础，选读解析我国古典文献中蕴含的先进设计思想，从我国优秀的传统文化中汲取设计灵感，增强对母语及母语文化的自豪感，解读杰出设计师们的经典设计范例及其设计文化思想、设计方法和设计流程。

  设计学科具有博采众长，体现中西方文化内涵、审美意象、艺术与科技结合的专业特点，双语教学不愧为一种有效启发思维的教学方法。曾有一位设计学资深教授对笔者说："双语教学就是'双倍'工作量的设计教学。"虽然，这是一句宽慰的话，但却说出了双语授课教师们的心声，因为，对于双语任课教师来说，既要完善中文的教学内容，又要准备已有教学内容的英文翻译，工作量之大难以想象，也是学生们难以体会到的。然而，有了适合的双语教材，即可落实双语课程。本教程的内容可用于《中外设计文化》、《设计文献研读》、《中外设计文献比较》和《设计文化概论》等课程的双语教学，均以中英对照的方式呈现、突出要点、解读详备、图文并茂，可以对照查阅。这是一部着眼于设计文化双语理解与创生设计构思的"启迪式"专业教材。

  本教程选取经典出众的设计范例，为书中所用范例提供作品的设计师有：任天、吕勤智、陈炜、朱昱宁、俞雪莱、郑昱。在此致以深切的谢忱。

应宜文

2017 年 8 月

本书课件 PPT 具体使用方法如下：

一、移动设备用户

1. 刮开图书封底网上增值服务标涂层，扫描二维码，按提示下载并安装我社"建工社学习"APP。

2. 在"建工社学习"APP 中登录后（已注册过我社网站的用户可直接登陆，未注册用户需登陆"中国建筑出版在线 www.cabplink.com"网站进行注册），再次扫描图书封底的二维码进行绑定，即可使用。每一个二维码只能绑定一次。

3. 点击"建工社学习"中的图书封面，即可进入观看相应资源。

二、计算机用户

1. 访问"ltjc.cabplink.com"网站，注册用户并登录，然后按照提示输入封底网上增值服务标涂层下的 ID 及 SN 进行绑定，每一组号码只能绑定一次。

2. 绑定成功后，即可进入观看相应资源。

如输入 ID 及 SN 号后无法通过验证，请及时与我社联系：

联系电话：4008-188-688；010-58934837（周一至周五工作时间）。